EARTH MATERIALS AND HEALTH

RESEARCH PRIORITIES FOR EARTH SCIENCE AND PUBLIC HEALTH

Committee on Research Priorities for Earth Science and Public Health

Board on Earth Sciences and Resources
Division on Earth and Life Studies

Board on Health Sciences Policy
Institute of Medicine

NATIONAL RESEARCH COUNCIL *AND*
INSTITUTE OF MEDICINE
OF THE NATIONAL ACADEMIES

THE NATIONAL ACADEMIES PRESS
Washington, D.C.
www.nap.edu

THE NATIONAL ACADEMIES PRESS • 500 Fifth Street, N.W. • Washington, DC 20001

 Supported by the National Science Foundation under Award No. 0106060; the U.S. Geological Survey, Department of the Interior, under Award No. 01HQAG0216; and the National Aeronautics and Space Administration under Award No. NNS04AA14G.

International Standard Book Number-13: 978-0-309-10470-8 (Book)
International Standard Book Number-10: 0-309-10470-X (Book)
International Standard Book Number-13: 978-0-309-66852-1 (PDF)
International Standard Book Number-10: 0-309-66852-2 (PDF)
Library of Congress Control Number: 2007921888

Additional copies of this report are available from the National Academies Press, 500 Fifth Street, N.W., Lockbox 285, Washington, DC 20055; (800) 624-6242 or (202) 334-3313 (in the Washington metropolitan area); Internet, http://www.nap.edu.

Cover: Design by Michele de la Menardiere. The *top right* is an image illustrating successful models of blood clotting (image courtesy of Nicole Rager-Fuller, National Science Foundation). The *top left* image is a high resolution photo of fluorite (image courtesy of U.S. Geological Survey; image source, AGI Image Bank, http://www.earthscienceworld.org/images).

Printed in the United States of America

THE NATIONAL ACADEMIES

Advisers to the Nation on Science, Engineering, and Medicine

The **National Academy of Sciences** is a private, nonprofit, self-perpetuating society of distinguished scholars engaged in scientific and engineering research, dedicated to the furtherance of science and technology and to their use for the general welfare. Upon the authority of the charter granted to it by the Congress in 1863, the Academy has a mandate that requires it to advise the federal government on scientific and technical matters. Dr. Ralph J. Cicerone is president of the National Academy of Sciences.

The **National Academy of Engineering** was established in 1964, under the charter of the National Academy of Sciences, as a parallel organization of outstanding engineers. It is autonomous in its administration and in the selection of its members, sharing with the National Academy of Sciences the responsibility for advising the federal government. The National Academy of Engineering also sponsors engineering programs aimed at meeting national needs, encourages education and research, and recognizes the superior achievements of engineers. Dr. Wm. A. Wulf is president of the National Academy of Engineering.

The **Institute of Medicine** was established in 1970 by the National Academy of Sciences to secure the services of eminent members of appropriate professions in the examination of policy matters pertaining to the health of the public. The Institute acts under the responsibility given to the National Academy of Sciences by its congressional charter to be an adviser to the federal government and, upon its own initiative, to identify issues of medical care, research, and education. Dr. Harvey V. Fineberg is president of the Institute of Medicine.

The **National Research Council** was organized by the National Academy of Sciences in 1916 to associate the broad community of science and technology with the Academy's purposes of furthering knowledge and advising the federal government. Functioning in accordance with general policies determined by the Academy, the Council has become the principal operating agency of both the National Academy of Sciences and the National Academy of Engineering in providing services to the government, the public, and the scientific and engineering communities. The Council is administered jointly by both Academies and the Institute of Medicine. Dr. Ralph J. Cicerone and Dr. Wm. A. Wulf are chair and vice chair, respectively, of the National Research Council.

www.national-academies.org

COMMITTEE ON RESEARCH PRIORITIES FOR EARTH SCIENCE AND PUBLIC HEALTH

H. CATHERINE W. SKINNER, *Chair*, Yale University, New Haven, Connecticut
HERBERT E. ALLEN, University of Delaware, Newark
JEAN M. BAHR, University of Wisconsin, Madison
PHILIP C. BENNETT, University of Texas, Austin
KENNETH P. CANTOR, National Cancer Institute, Bethesda, Maryland
JOSÉ A. CENTENO, Armed Forces Institute of Pathology, Washington, D.C.
LOIS K. COHEN, National Institute of Dental and Craniofacial Research, Bethesda, Maryland
PAUL R. EPSTEIN, Harvard Medical School, Boston, Massachusetts
W. GARY ERNST, Stanford University, California
SHELLEY A. HEARNE, Trust for America's Health, Washington, D.C.
JONATHAN D. MAYER, University of Washington, Seattle
JONATHAN PATZ, University of Wisconsin, Madison
IAN L. PEPPER, University of Arizona, Tucson

Liaison from the Board on Health Sciences Policy

BERNARD D. GOLDSTEIN, University of Pittsburgh, Pennsylvania

National Research Council Staff

DAVID A. FEARY, Study Director (Board on Earth Sciences and Resources)
CHRISTINE M. COUSSENS, Program Officer (Board on Health Sciences Policy)
JENNIFER T. ESTEP, Financial Associate
CAETLIN M. OFIESH, Research Associate
AMANDA M. ROBERTS, Senior Project Assistant (until August 2006)
NICHOLAS D. ROGERS, Senior Project Assistant (from September 2006)

BOARD ON EARTH SCIENCES AND RESOURCES

National Research Council Staff

BOARD ON HEALTH SCIENCE POLICY

Preface

We live in an era with unparalleled opportunities to practice disease prevention based on knowledge of the earth environment. Although globally distributed early warning systems can monitor physical hazards such as earthquakes and tsunamis, chemical hazards on the other hand—whether actual or potential and natural or anthropogenically induced—remain difficult to accurately identify in time and space. Such hazards often have lengthy asymptomatic latency periods before disability or disease becomes evident. The scientific information available from the earth sciences—knowledge about earth materials and earth processes, the normal environment, or potential hazards—is essential for the design and maintenance of livable environments and a fundamental component of public health.

A global perspective is necessary when considering the interlinked geochemical and biochemical research issues at the intersection of the earth sciences and public health. The air that carries viruses or earth-sourced particulate matter is clearly global and circulates beyond human control. Pathogens in soil and water have enhanced potential for global spread as food is increasingly transported worldwide. And the availability of irrigation and potable water is increasingly acknowledged as a worldwide issue. As the United Nations International Year of Planet Earth (2008) approaches, it is particularly gratifying that "Earth and Health: Building a Safer Environment" is one of the 10 research themes. This presents an important opportunity for the earth science and public health research communities on a global scale; the committee hopes that this report will provide research focal points and suggest mechanisms to improve communication and collaboration between these communities.

The broad purview of the committee's task has been a blessing rather than a curse. As the topics and issues addressed by the committee ranged from global to personal, remarkable opportunities arose for interaction among committee members from diverse backgrounds and with differing scientific vocabularies and knowledge bases. From the immense range of potential research opportunities, the committee members were able to achieve a consensus on the priority research directions and mechanisms that we believe will contribute to improved public health and better safeguarding of our earth environment.

H. Catherine W. Skinner,
Chair

Acknowledgments

This report was greatly enhanced by input from participants at the workshop and public committee meetings held as part of this study: Ludmilla Aristilde, E. Scott Bair, Anthony R. Berger, Gordon E. Brown, Jr., Herbert T. Buxton, Margaret Cavanaugh, Rachael Craig, Ellen Marie Douglas, Barbara L. Dutrow, Jonathan E. Ericson, Rodney C. Ewing, Robert B. Finkelman, Charles P. Gerba, Charles G. Groat, Linda C.S. Gundersen, Mickey Gunter, Stephen C. Guptill, John A. Haynes, Richard J. Jackson, Michael Jerrett, K. Bruce Jones, Ann Marie Kimball, P. Patrick Leahy, Louise S. Maranda, Perry L. McCarty, Catherine Pham, Geoffrey S. Plumlee, Donald Rice, Joshua P. Rosenthal, Carol H. Rubin, Harold H. Sandstead, Samuel M. Scheiner, Ellen K. Silbergeld, Barry Smith, Alan T. Stone, Lesley A. Warren, Robert T. Watson, Samuel H. Wilson, Scott D. Wright, Harold Zenick, and Herman Zimmerman. These presentations and discussions helped set the stage for the committee's fruitful discussions in the sessions that followed.

This report has been reviewed in draft form by individuals chosen for their diverse perspectives and technical expertise, in accordance with procedures approved by the National Research Council's (NRC) Report Review Committee. The purpose of this independent review is to provide candid and critical comments that will assist the institution in making its published report as sound as possible and to ensure that the report meets institutional standards for objectivity, evidence, and responsiveness to the study charge. The review comments and draft manuscript remain confidential to protect the integrity of the deliberative process. We wish to thank the following individuals for their participation in the review of this report:

John C. Bailar III, Department of Health Studies, The University of Chicago (emeritus), Washington, D.C.
Thomas A. Burke, Department of Health Policy and Management, Johns Hopkins University, Bloomberg School of Public Health, Baltimore, Maryland
Kristie L. Ebi, Health Sciences Practice, Exponent, Alexandria, Virginia
Rodney Klassen, Applied Geochemistry, Geological Survey of Canada, Ottawa
Ben A. Klinck, Chemical and Biological Hazards Programme, British Geological Survey, Keyworth, Nottingham, United Kingdom
Jonathan M. Samet, Department of Epidemiology, Johns Hopkins University, Bloomberg School of Public Health, Baltimore, Maryland
Rien van Genuchten, Agricultural Research Service, U.S. Department of Agriculture, Riverside, California
Philip Weinstein, School of Population Health, University of Western Australia, Crawley

Although the reviewers listed above provided many constructive comments and suggestions, they were not asked to endorse the conclusions or recommendations nor did they see the final draft of the report before its release. The review of this report was overseen by David S. Kosson, Department of Civil and Environmental Engineering, Vanderbilt University, Nashville, Tennessee, and Edward B. Perrin, School of Public Health, University of Washington, Seattle. Appointed by the NRC, they were responsible for making certain that an independent examination of this report was carried out in accordance with institutional procedures and that all review comments were carefully considered. Responsibility for the final content of this report rests entirely with the authoring committee and the institution.

Contents

Summary

The interactions between earth materials and processes and human health are pervasive and complex. In some instances, the association between earth materials and disease is clear—certain fibrous (asbestos) minerals and mesothelioma, radon gas and lung cancer, dissolved arsenic and a range of cancers, earthquakes and physical trauma, fluoride and dental health—but these instances are overshadowed by the many cases where individual earth components, or more commonly mixtures of earth materials, are suspected to have deleterious or beneficial health impacts. Unraveling these more subtle associations will require substantial and creative collaboration between earth and health scientists.

The surge of interest and research activity investigating relationships between public health and earth's environment that commenced in the 1960s has not been sustained. Today, few researchers span the interdisciplinary divide between the earth and public health sciences, and "stovepipe" funding from agencies provides little incentive for researchers to reach across that divide. The limited extent of interdisciplinary cooperation has restricted the ability of scientists and public health workers to solve a range of complex environmental health problems, with the result that the considerable potential for increased knowledge at the interface of earth science and public health has been only partially realized and opportunities for improved population health have been threatened.

In response to this situation, the National Science Foundation, U.S. Geological Survey, and National Aeronautics and Space Administration requested that the National Research Council undertake a study to explore avenues for interdisciplinary research that would further knowledge at the interface between the earth science and public health disciplines. The study committee was charged to advise on the high-priority research activities that should be undertaken for optimum societal benefit, to describe the most profitable areas for communication and collaboration between the earth science and public health communities, and to respond to specific tasks:

- Describe the present state of knowledge in the emerging medical geology field.
- Describe the connections between earth science and public health, addressing both positive and negative societal impacts over the full range from large-scale interactions to microscale biogeochemical processes.
- Evaluate the need for specific support for medical geology research and identify any basic research needs in bioscience and geoscience required to support medical geology research.
- Identify mechanisms for enhanced collaboration between the earth science and medical/public health communities.
- Suggest how future efforts should be directed to anticipate and respond to public health needs and threats, particularly as a consequence of environmental change.

RESEARCH PRIORITIES

The committee addressed this charge by focusing its analysis on human exposure pathways—what we breathe, what we drink, what we eat, and our interactions with earth materials through natural and anthropogenic earth perturbations (e.g., natural disasters, land cover modifications, natural resource use). Specific examples for each exposure pathway are presented to highlight the state of existing knowledge, before listing priority collaborative research activities for each exposure pathway. These research activities are grouped into three broad crosscutting themes: (1) improved understanding of the source, fate, transport, bioavailability, and impact of potentially hazardous or beneficial earth materials; (2) improved risk-based hazard mitigation, based on improved understanding of the public health effects of natural hazards under existing and future climatic regimes; and (3) research to understand the health risks arising from disturbance of terrestrial systems as the basis for prevention of new exposures. The committee received suggestions for broad research initiatives and specific research activities from national and international participants from the earth science, public health, and government funding communities at an open workshop, and these suggestions formed the basis for deliberations to identify the research themes considered by the committee to have the highest priority. In compiling these recommendations the committee required that the research proposed must involve collaboration between researchers from both the earth science and the public health communities and did not consider the abundant examples of valuable research that could be undertaken primarily within one or other of the disciplines.

Earth Material Exposure Assessments—Understanding Fate and Transport

Assessment of human exposure to hazards in the environment is often the weakest link in most human health risk assessments. The physical, chemical, and biological processes that create, modify, or alter the transport and bioavailability of natural or anthropogenically generated earth materials remain difficult to quantify, and a vastly improved understanding of the spatial and geochemical attributes of potentially deleterious earth materials is a critical requirement for effective and efficient mitigation of the risk posed by such materials. An improved understanding of the source, fate, rate and transport, and bioavailability of potentially hazardous earth materials is an important research priority. Collaborative research should include:

- Addressing the range of issues associated with airborne *mixtures* of pathogens and physical and chemical irritants. The anticipation and prevention of health effects caused by earth-sourced air pollution prior to the onset of illness requires quantitative knowledge of the geospatial context of earth materials and related disease vectors.
- Determining the influence of biogeochemical cycling of trace elements in water and soils as it relates to low-dose chronic exposure via toxic elements in foods and ultimately its influence on human health.
- Determining the distribution, survival, and transfer of plant and human pathogens through soil with respect to the geological framework.
- Improving our understanding of the relationship between disease and both metal speciation and metal-metal interaction.
- Identifying and quantifying the health risks posed by "emerging" contaminants, including newly discovered pathogens and pharmaceutical chemicals that are transported by earth processes.

Improved Risk-Based Hazard Mitigation

Natural earth processes—including earthquakes, landslides, tsunamis, and volcanoes—continue to cause numerous deaths and immense suffering worldwide. As climates change, the nature and distribution of such natural disasters will undoubtedly also change. Improved risk-based hazard mitigation, based on improved understanding of the public health effects of natural hazards under existing and future climatic regimes, is an important research priority. Such collaborative research should include:

- Determining processes and techniques to integrate the wealth of information provided by the diverse earth science, engineering, emergency response, and public health disciplines so that more sophisticated scenarios can be developed to ultimately form the basis for improved natural hazard mitigation strategies.

Assessment of Health Risks Resulting from Human Modification of Terrestrial Systems

Human disturbances of natural terrestrial systems—for example, by activities as diverse as underground resource extraction, waste disposal, or landcover and habitat change—are creating new types of health risks. Research to understand and document the health risks arising from disturbance of terrestrial systems is a critical requirement for alleviating existing health threats and preventing new exposures. Such collaborative research should include:

- Analysis of the effect of geomorphic and hydrological landsurface alteration on disease ecology, including emergence/resurgence and transmission of disease.
- Determining the health effects associated with water quality changes induced by novel technologies and other strategies currently being implemented, or planned, for extending groundwater and surface water supplies to meet increasing demands for water by a growing world population.

PROMOTING COLLABORATION

Geospatial information—geological maps for earth scientists and epidemiological data for public health professionals—is an essential integrative tool that is fundamental to the activities of both communities. The application of modern complex spatial analytical techniques has the potential to provide a rigorous base for integrated earth science and public health research by facilitating the analysis of spatial relationships between public health effects and natural earth materials and processes. Research activity should be focused on the development of high-resolution, spatially and temporally accurate models for predicting disease distribution that incorporate layers of geological, geographic, and socioeconomic data.

This will require development of improved technologies for high-resolution data generation and display.

Before it will be possible to take advantage of the considerable power of modern spatial analysis techniques, a number of issues associated with data access will need to be addressed. Improved coordination between agencies that collect health data will be required, and health data from the different agencies and offices will need to be merged and made available in formats that are compatible with GIScience analysis. Existing restrictions on obtaining geographically specific health data, while important for maintaining privacy, severely inhibit effective predictive and causal analysis. To address this, it will be necessary for all data collected by federal, state, and county agencies to be geocoded and geographically referenced to the finest scale possible, and artificial barriers to spatial analysis resulting from privacy concerns need to be modified to ensure that the enormous power of modern spatial analysis techniques can be applied to public health issues without affecting privacy.

Although important gains have been made *within* individual funding agencies toward support for interdisciplinary research, a dearth of collaboration and funding *between* agencies has restricted significant scientific discovery at the interface of public health and earth science. The committee suggests that, for there to be substantial and systemic advances in interdisciplinary interaction, a formal multiagency collaboration support system needs to replace the existing ad hoc nature of collaborations and funding support. Despite wariness about proposing yet another bureaucratic structure, the committee believes that a multitiered hierarchical management and coordination mechanism could provide a structure by which the relevant funding agencies would be encouraged to interact for improved communication and collaboration.

The synergies from interdisciplinary interactions provide the basis for innovative and exciting research that can lead to new discoveries and greater knowledge. As both the researchers, and the agencies that fund their research, seek to increase support for interdisciplinary research, the time is right for the earth science and public health communities to take advantage of the opportunity to promote true collaboration—there is no doubt that society will ultimately derive significant health benefits from the increased knowledge that will derive from research alliances.

The interface between the earth sciences and public health is pervasive and enormously complex. Collaborative research at this interface is in its infancy, with great potential to ameliorate the adverse health effects and enhance the beneficial health effects from earth materials and earth processes. The earth science and public health research communities share a responsibility and obligation to work together to realize the considerable potential for both short-term and long-term positive health impacts.

Section I

Introduction

1

Introduction

The nature and extent of our interactions with the natural environment have a profound impact on human well-being. Earth science includes the broad subdisciplines of geology, geophysics, geochemistry, geomorphology, soil science, hydrology, mineralogy, remote sensing, mapping, climatology, volcanology, physical geography, and seismology. As such, earth science describes a substantial component of this natural environment, encompassing the key terrestrial materials, associations, and processes that have both beneficial and adverse impacts on public health. Despite this association between public health and the natural environment, geologists, geophysicists, and geochemists have intensively studied the earth for the past two centuries with only passing appreciation for the impacts of the geological substrate, earth materials, and earth processes on human health. Similarly, although health scientists have a rapidly expanding understanding of individual physiology and the epidemiology of human populations on local to global scales, most modern public health practitioners have only limited awareness of the extent to which the earth environment impinges on public health.

Although valuable linkages do currently exist between the earth science and public health communities, the limited extent of interdisciplinary cooperation has restricted the ability of scientists and public health workers to solve a range of complex environmental health problems, with the result that the considerable potential for increased knowledge at the interface of earth science and public health has been only partially realized. The linkage of earth science and public health is not about the relevance of earth science knowledge to health, or vice versa—rather, the issue addressed here is the generally inadequate appreciation of the potential benefits of this interface and the consequent diminished priority that it is accorded.

NEED FOR COLLABORATIVE RESEARCH

Historically, it was known that some geographic locations were associated with specific diseases in humans and animals. Marco Polo recognized hoof diseases in animals that had consumed certain plants (later determined to be selenium-accumulating plants) and observed physical abnormalities (goiters) that he attributed to the local water supply. Recognition of the role of iodine to alleviate goiter emphasizes the importance of research at the interface of earth science and public health, and in fact iodine deficiency is one of the single most preventable causes of mental retardation (Delange et al., 2001). Similarly, the addition of fluoride to drinking water and toothpaste, based on recognition of the beneficial effects of naturally fluoridated water, has been hailed as one of the top 10 public health achievements of the twentieth century (CDC, 1999). For communities of more than 20,000 people, the cost savings from prevention of dental cavities as a result of water fluoridation has been estimated as 38 times the cost of fluoride addition (Griffin et al., 2001).

Such instances are far outweighed by examples where prior knowledge of earth science and improved understanding of the characteristics of earth materials could have informed the decision-making process and prevented disease. Volcanic aerosols, gases and ash, airborne and waterborne fibrous minerals, and toxic metals in soils and plants are all examples presented later in this report where earth materials have adversely affected human health (see Box 1.1).

In a series of reports more than 20 years ago, National Research Council (NRC) committees described the contemporary understanding of interactions between earth's geochemical environment and public health (NRC, 1974, 1977, 1978, 1979, 1981). This report presents a broad update describing our understanding of the interactions between earth materials and public health, provides an introduction to successful past cooperative scientific activities at the interface of the earth and health sciences, and suggests future avenues for crossover and integration of research for the common good of humankind.

COMMITTEE CHARGE AND SCOPE OF THE STUDY

Recognizing the current disconnect between research carried out by the earth science and public health communities, the National Science Foundation (NSF), U.S. Geological Survey (USGS), and National Aero-

BOX 1.1
Arsenic Contamination of Groundwater in Bangladesh

One of the clearest examples of the crossover between the earth sciences and public health is the infamous problem of arsenic in Bangladesh and West Bengal, India. In the 1970s, the United Nations Educational, Scientific and Cultural Organization (UNESCO) funded the digging of simple tube wells (up to 150 m deep) into rapidly deposited, unconsolidated deltaic sediments of the Ganges and Brahmaputra river systems. The goal was to minimize the use of surface waters for domestic use and thereby reduce the devastating effects of cholera and diarrhea, which were responsible for many deaths among the young and the elderly. Water from tube wells, it was thought, would replace the seriously contaminated surface water supply with adequate fresh, pure groundwater. The deltaic sediments, consisting chiefly of mud, silt, and sand, also contain organic matter and trace minerals carried from the upper reaches of the river systems. The iron oxides in these sediments are effective at scavenging arsenic (and other oxyanions such as phosphate), and when the iron oxides are reduced by iron-reducing bacteria (reductive dissolution), the associated ions such as arsenate are mobilized. Consumption and crop irrigation of arsenic-bearing water, in some cases with arsenic contents greater than 500 $\mu g\ L^{-1}$, resulted in widespread arsenic poisoning which was especially prevalent in people at high risk due to poor nutrition. The result was a horrible disease most commonly manifested by skin lesions and cancer. Although the effect of the arsenic varies with the element species (As^{3+}, As^{5+}), it mostly acts through inactivation of enzyme systems, with trivalent arsenic being the most injurious (Ginsburg, and Lotspeich, 1963). In this region, more than 30 million people are drinking water that contains arsenic at concentrations exceeding the Bangladesh drinking water guidance value of 0.05 mg L^{-1} (i.e., 50 $\mu g\ L^{-1}$) (Rahman et al., 2003), and the number would be considerably greater if the World Health Organization–recommended guideline value of 10 $\mu g\ L^{-1}$ (WHO, 2001) were used. Further, at least 175,000 people have skin lesions caused by arsenic poisoning.[1]

---

[1] As this report was being readied for printing, the National Academy of Engineering announced that Dr. Abul Hussam, from George Mason University, had been awarded the 2007 Grainger Challenge Prize for Sustainability Gold Award for developing a household water treatment system to remove arsenic from drinking water in Bangladesh.

BOX 1.2
Statement of Task

A National Research Council ad hoc committee will assess the present status of research at the interface between medicine and earth science, and will advise on the high priority research activities that should be undertaken for optimum societal benefit. The committee will report on the most profitable areas for communication and collaboration between the earth science and medical communities, recognizing both the infectious disease and environmental components. The committee is specifically tasked to:

- Describe the present state of knowledge in the emerging medical geology field.
- Describe the connections between earth science and public health, addressing both positive and negative societal impacts over the full range from large-scale interactions to microscale biogeochemical processes.
- Evaluate the need for specific support for medical geology research, and identify any basic research needs in bioscience and geoscience required to support medical geology research.
- Identify mechanisms for enhanced collaboration between the earth science and medical/public health communities.
- Suggest how future efforts should be directed to anticipate and respond to public health needs and threats, particularly as a consequence of environmental change.

nautics and Space Administration (NASA) requested that the NRC undertake a study to explore avenues for interdisciplinary research that would further knowledge at the interface between these disciplines (see Box 1.2). The ultimate goal is to encourage collaboration to comprehensively address human health problems in the context of the geological environment.

The committee assembled by the National Academies to address this task held three open, information-gathering meetings, where representatives from federal and state agencies, the academic community, and professional societies provided information and perspectives on the committee's task. One of these meetings was a three-day workshop, where a combination of presentations and breakout groups allowed for extensive interdisciplinary exchange of data and concepts. During closed sessions, the committee deliberated on the broad issues involved in the integration of disparate disciplinary approaches. Although recognizing that processes linking the solid earth with the biosphere, the oceans, and the atmosphere represent a continuum, the committee concentrated its attention on the relatively direct geological drivers of human health. Recent NRC reports have described interactions between human health issues and the oceans (NRC, 1999d) and between human health and the atmosphere (NRC, 2001a); accordingly, this committee focused on the continental environment and only considered oceanic and the atmospheric effects on human health through their interactions with on-land geology (e.g., volcanic emanations, particulate matter). The committee excluded the extremely important domain of agriculture as beyond its purview.

The recent publication of two major texts on the earth sciences and health (Skinner and Berger, 2003; Selinus et al., 2005) reflects the increased attention being focused by researchers on important interactions between these fields. Together, these books provide a comprehensive description of current understanding of the relationship between the natural environment and public health, as well as numerous examples describing the connections and interactions between these fields. Rather than attempt to cover the same material, the NRC committee sought to build on these works by focusing its endeavors on understanding the vast array of potential research directions at the interface of earth science and public health and to identify those that it considers to have the highest priority.

The process of identifying the priority research areas presented in this report was based on the discussions and conclusions at the open workshop hosted by the committee. The four workshop breakout groups—in each case coordinated by a committee member and including members of the earth science, public health, and governmental communities—reported back with recommendations that described important research areas for the committee's consideration. The committee focused on those areas that required full collaboration by both earth science and public health researchers and did not consider the numerous examples of valuable research topics that could be undertaken primarily within one of these research disciplines without requiring significant participation by the other.

When members of two distinct professional communities who have traditionally had little interaction come together on a study committee such as this, it is not surprising that issues of vocabulary and definition rapidly emerge. Recognizing that acceptance of the recommendations contained here by both communities will, to some extent, depend on both being able to easily understand the presentation of the ideas and concepts without either feeling that there is an overlay of technical jargon, the committee has attempted to ensure that such jargon is kept to a minimum throughout the report. In some cases, this has resulted in ideas, concepts, and situations being presented in a somewhat simplistic manner; nevertheless, the committee considers that such simplicity is essential. This approach is reflected in Chapter 2, where basic earth science concepts are presented for the public health community and basic human physiological concepts are presented for the earth science community.

The committee decided that the report could best highlight the state of knowledge by focusing on the interactions between earth science and public health through the public health reference frame—that is, through human exposure routes. The committee organized these sections of the report into what we breathe (Chapter 3), what we drink (Chapter 4), and what we eat (Chapter 5). Public health interactions with earth perturbations, both natural (e.g., earthquakes, volcanic eruptions) and anthropogenic (e.g., extractive industries), are described in Chapter 6. In these separate sections, the committee employs specific examples to focus on, and highlight, the state of present knowledge. These considerations of cause and health effect resulting from exposure to earth materials give rise to a range of research priorities for each exposure pathway—in each case, those that have been identified by the committee require active collaboration between researchers from both the earth science and the public health communities. The role of geospatial information—geological maps for earth scientists and epidemiological data for public health professionals—is recognized as an essential integrative tool that is fundamental to the activities of both communities (Chapter 7), and a number of suggestions are presented for mechanisms to promote and enhance collaboration (Chapter 8). Finally, the committee presents a series of conclusions and recommendations, based on the opportunities for research collaboration described in Chapters 3 through 7, which are designed to enhance integration of the earth and public health sciences (Chapter 9).

2

Earth Processes and Human Physiology

Except for radiant energy from the sun, the resources necessary for sustaining life are derived chiefly from the near-surface portions of the land, sea, and air. Intensive utilization of earth materials has enhanced the quality of human life, especially in the developed nations. However, natural background properties and earth processes such as volcanic eruptions, as well as human activities involving the extraction, refining, and manufacturing of mineral commodities, have led to unwanted side effects such as environmental degradation and health hazards. Among the latter are airborne dusts and gases, chemical pollutants in agricultural, industrial, and residential waters, and toxic chemical species in foodstuffs and manufactured products. Of course, at appropriate levels of ingestion and assimilation, most earth materials are necessary for life, but underdoses and overdoses have adverse effects on human health and aging.

Although the environmental concentration of a substance is important and relatively easy to measure, its specific chemical form (a function of the biogeochemical environment, complex species interactions, Eh, and pH) determines the substance's reactivity and therefore its bioaccessibility. In the case of earth materials, specific mineralogical characteristics (e.g., mineralogy, grain size) must also be considered together with these chemical factors when assessing bioaccessibility. Thus, a number of analytical measurements are required to accurately assess the bioavailability of a naturally occurring chemical and mineralogical species. For most, an optimal dose range enhances health, whereas too little (deficiency) or too much (toxicity) have adverse impacts. Because the bioavailabilities of a spectrum of earth materials present in the environment constitute critical variables that influence human health—particularly where regional and local "hotspots" of earth material deficiency or toxicity occur—the bioavailabilities of earth materials must be quantified by collaborative, integrated geological and biomedical research. To understand the physiological responses of the human body to the ingestion and assimilation of earth materials, this chapter begins by briefly describing the dynamic geological processes responsible for the areal disposition of earth materials in the near-surface environment, with particular attention to soil characteristics. This is followed by a brief description of those aspects of human physiology that are—through their responses to bioaccessible nutrients and hazardous materials—directly responsive to the biogeochemical environment.

EARTH PROCESSES

The near-surface portions of the planet and their complex couplings with—and feedbacks from—the atmosphere, hydrosphere, and biosphere make up the interactive earth systems so crucial for life. In turn, these dynamic systems are a reflection of the origin and geological evolution of the earth in the context of solar system formation. The following brief review of earth's deep-seated and surficial processes provides the physical context for the public health component of human interactions with the earth.

Planetary Architecture and Crustal Dynamics

The solid earth consists of a series of nested shells. The outermost thin skin, or crust, overlies a magnesium silicate-rich mantle, the largest mass of the planet. Beneath the mantle is the earth's iron-nickel core. The terrestrial surface is unique among the planets of our solar system, possessing an atmosphere, global oceans, and both continents and ocean basins. Incoming sunlight powers oceanic-atmospheric circulation. Solar energy absorbance and transfer mechanisms are responsible for the terrestrial climate and its variations, as well as for cyclonic storms and coastal flooding. In the solid portions of the planet, the escape of buried heat through mantle flow has produced the earth's crust, as well as energy and mineral deposits and all terrestrial substances necessary for life in the biosphere.

Although imperceptible to humans without geophysical monitoring, continuous differential vertical and horizontal motions characterize the earth's crust. This remarkable mobility explains the growth and persistence of long-lived, high-standing continents and the relative youth of low-lying ocean basins, although the former are being planed down by erosion and the latter are being filled through sedimentary deposition. The earth's surface is continuously being reworked, and a dynamic equilibrium has been established between competing agents of crustal erosion and deposition (external processes) *versus* crustal construction (an internal process). Crustal deformation, a consequence of mantle dynamics, is the ultimate cause of many geological hazards, including earthquakes and tsunamis, volcanic eruptions, and landslides. In addition to the direct fatalities and injuries, natural catastrophes result in the displacement of surviving populations into unhealthy environments where communicable diseases can—and often do—spread widely.

Plate Tectonics—Origins of Continental and Ocean Crust

Scientists have studied the on-land geology of the earth for more than two centuries, and much is known concerning the diverse origins of the continental crust, its structure, and constituent rocks and minerals (see Earth Materials below). Within the past 35 years, marine research has elucidated the bathymetry, structure, and physicochemical nature of the oceanic crust, and as a result we have a considerably improved appreciation of the manner in which various parts of the earth have evolved with time. A startling product of this work was the realization that, beneath the relatively stiff outer rind of the planet (the lithosphere), portions of the more ductile mantle (the asthenosphere) are slowly flowing. Both continental and oceanic crusts form only the uppermost, near-surface layers of great lithospheric plates; differential motions of these plates—plate tectonics—are coupled to the circulation of the underlying asthenosphere on which they rest. The eastern and western hemispheric continents are presently drifting apart across the Atlantic Ocean and have been doing so for more than 120 million to 190 million years. Locally, continental fragments came together in the past and others are presently colliding, especially around the Pacific Rim.

Mid-ocean ridges represent the near-surface expression of hot, slowly ascending mantle currents with velocities on the order of a few centimeters per year. Whether this upwelling is due to part of a convection cell that returns asthenosphere to shallower levels after it has been dragged to depths by a lithospheric plate sinking elsewhere, or is a consequence of deeply buried thermal anomalies that heat and buoy up the asthenosphere, is not known, but both processes probably occur to varying degrees. Approaching the seafloor, the rising mantle undergoes decompression and partial melting to generate basaltic liquid. The magma within the upwelling asthenosphere is less dense and thus even more buoyant. It rises toward the interface with seawater and solidifies to form the oceanic crust, capping the stiffer, less buoyant mantle. The mid-oceanic ridges—

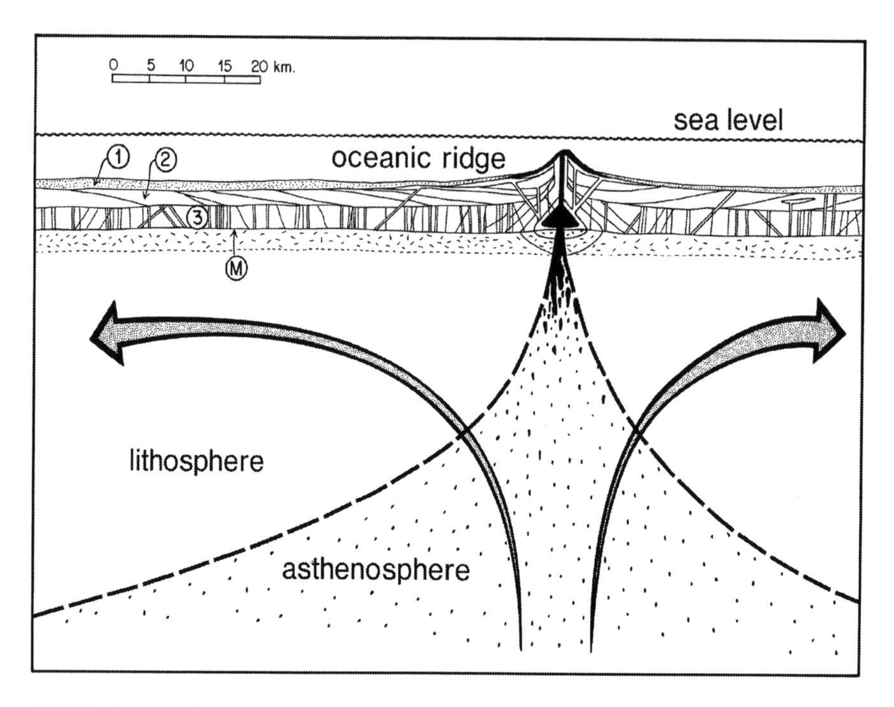

FIGURE 2.1 Schematic cross-section of a mid-oceanic ridge spreading center (a divergent plate boundary). Curvilinear mantle flow lines (arrows) show the circulation paths followed by rising asthenosphere and its cooling and conversion to lithosphere. Basaltic magma is shown as black coalescing blobs. Layers 1, 2, and 3 are deep-sea sediments, basaltic lava flows, and intrusive equivalents, respectively. M marks the Mohorovicic Discontinuity (the crust-mantle boundary).
SOURCE: Ernst (1990).

divergent plate boundaries—are spreading centers where the cooling lithospheric plates that overlie the ductily flowing mantle currents are transported at right angles away from the ridge (see Figure 2.1).

As it moves away from the ridge axis, the cooling oceanic lithosphere gradually thickens at the expense of the upper part of the asthenosphere. Heat is continuously lost, so the lithosphere-asthenosphere boundary (solid, rigid mantle above; incipiently molten, ductile mantle below), which is very close to the sea bottom beneath the oceanic ridge, descends to greater water depths away from the spreading center because its overall density increases. Unlike light continental lithosphere floating on a denser mantle, the oceanic lithosphere has a slightly greater density than the asthenosphere below, and so the oceanic plate will sink back into the deep mantle where geometrically possible.

An oceanic plate moves away from the ridge axis until it reaches a convergent plate boundary. Here, one slab must return to the mantle to conserve volume—the process of subduction (see Figure 2.2). A bathymetric low, or trench, marks the region where bending of the down-going oceanic slab is greatest. It is difficult for continental crust-capped lithosphere to sink because it is less dense than the mantle below; however, due to the descent of oceanic lithosphere, the dragging of a segment of continental crust into and down the inclined subduction zone occasionally takes place.

Production of new oceanic crust along submarine ridge systems results from this plastic flow of the deep earth, as does addition to—and deformation of—the continental crust in the vicinity of seismically and volcanically active continental margins and island arcs. Oceanic ridges are sited over upwelling mantle columns, whereas along subduction zones, lithospheric plates are descending beneath active continental mar-

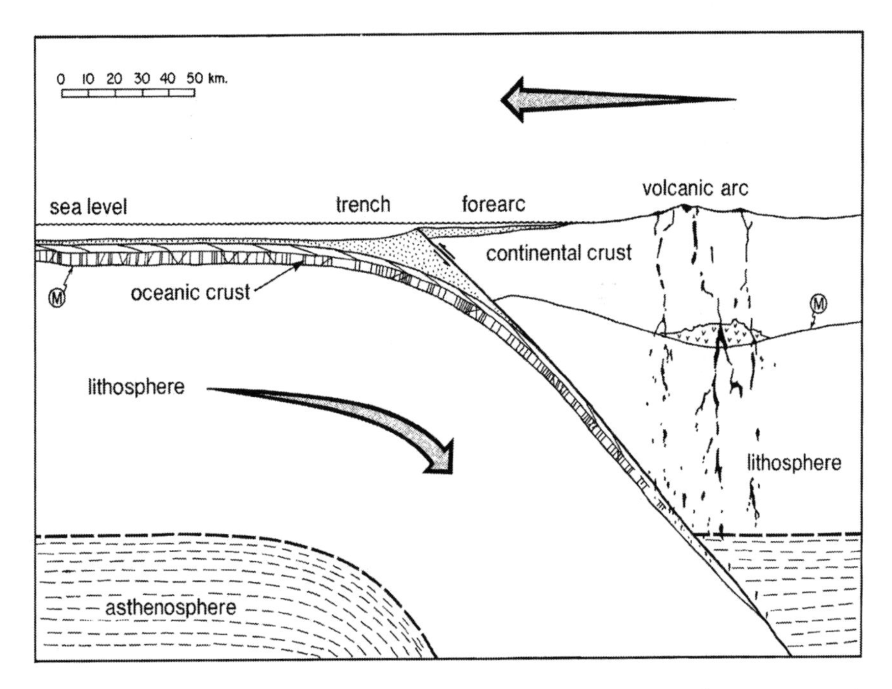

FIGURE 2.2 Schematic cross-section of a subduction zone, involving an oceanic trench and island arc-continental margin (a convergent plate boundary). Curvilinear mantle flow (arrow) shows the paths followed by cooling, descending lithosphere. Island arc magma is shown as black coalescing blobs. M is the Mohorovicic Discontinuity (the crust-mantle boundary).
SOURCE: Ernst (1990).

gins and island arcs. In contrast to submarine ridges, however, continents are also typified by mountain belts that display evidence of great crustal shortening and thickening (e.g., the Appalachians, Himalayas, and Alps). These compressional mountains contain great tracts of preexisting layered rocks, now contorted into fault-bounded blocks of folded rock. Such collisional mountain belts mark the sites of present or ancient plate boundaries.

Geological Catastrophes—Earthquakes, Volcanic Eruptions, and Landslides

Earthquakes are concentrated along plate boundaries, and earthquake locations in the oceans outline the edges of the rather young, approximately 50 km thick, homogeneous plates. In contrast, continental lithosphere may be as much as 200 to 300 km thick and is on average much older—up to 3.9 billion years old. As a consequence, these continent-capped plates consist of a diversity of rock types with variable strengths, being transected by numerous discontinuities and zones of weakness, which reflect a tortured history of repeated rifting, crustal amalgamation, and mountain building.

Where oceanic lithosphere descends beneath continents and island arcs along subduction zones, seismicity is situated at progressively greater depths farther inland as the stable lithospheric plate progressively overrides the sinking plate (see Figure 2.2). For this reason, active continental margins exhibit a broad zone of earthquakes. Seismicity around the Circumpacific is intense because oceanic lithosphere sinks beneath the continental edges of the Americas and Australasia. Large earthquakes—such as the Sumatran earthquake of December 26, 2004—are episodic, with intensities roughly proportional to the shaking time and the length of the crustal segment that ruptured in the specific seismic event.

Volcanism is a consequence of partial melting of the down-going lithospheric plate at depths approaching or exceeding 100 km. Such magmas rise buoyantly into the earth's crust in island arcs and continental margins where they form volcanic chains and subjacent batholiths. This is also the region where plate convergence and contraction builds structural mountain belts, resulting in crustal thickening, rugged topography, and high erosion rates—such belts are characterized by landslides, mudflows, and other mass movements.

Earth Materials

The earth's crust constitutes far less than 1 percent of the entire planetary mass but represents the nurturing substrate for virtually all life on land and much of the life in the oceans. The biosphere predominantly occupies the near-surface skin of the solid crust—the upper, illuminated portions of the oceans and the lowermost zones of the atmosphere. To investigate and quantify the human health and longevity effects due to the presence and bioassimilation of earth materials, we need to understand the nature of the constituents that make up the earth's crust—minerals and rocks.

A *mineral* is a naturally occurring, inorganically produced solid that possesses a characteristic chemistry or limited range of compositions, and a periodic, three-dimensional atomic order or polymerization (i.e., crystal structure). The diagnostic physical properties of a mineral, such as hardness, fracture, color, density, index of refraction, solubility, and melting temperature, are unique and specific consequences of a mineral's chemical constitution, bonding, and crystal structure. Important minerals that make up the near-surface crust, and their chemical formulas, include quartz, SiO_2; alkali feldspar, $(K, Na)AlSi_3O_8$; plagioclase feldspar, $(Na, Ca)Al_{1\text{-}2}Si_{3\text{-}2}O_8$; olivine, $(Mg, Fe)_2SiO_4$; garnet, $(Mg, Fe)_3Al_2Si_3O_8$; pyroxene, $(Mg, Fe)SiO_3$; amphibole, $Ca_2(Mg, Fe)_5Si_8O_{22}(OH)_2$; muscovite, $KAl_2Si_3AlO_{10}(OH)_2$; biotite, $K(Mg, Fe)_3Si_3AlO_{10}(OH)_2$; talc, $Mg_3Si_4O_{10}(OH)_2$; serpentine, $Mg_6Si_4O_5(OH)_8$; kaolin, $Al_4Si_4O_{10}(OH)_8$; calcite, $CaCO_3$; pyrite, FeS_2; and hematite, Fe_2O_3. Some of these minerals are produced deep within the earth, some form as weathering products at the earth's surface, and some are formed by biological processes (e.g., limestones containing the fossilized remains of marine organisms, formed by biomineralizing processes that are analogous to the human processes that form bones and teeth).

A *mineraloid* is a naturally occurring solid or liquid that lacks a rigorous, periodic atomic structure. The chemical compositions and physical properties of mineraloids range widely. Such substances are more weakly bonded than compositionally similar minerals; most behave like viscous fluids. Volcanic glass, amber, coal, and petroleum are examples of mineraloids.

A *rock* is a naturally occurring, cohesive, multigranular aggregate of one or more minerals and/or mineraloids, making up an important mapable part of the crust at some appropriate scale. The mineralogical and bulk compositions of a rock are a function of its origin. Geologists recognize three main rock-forming processes; hence, there are three principal classes of rocks:

1. *Igneous*—A molten, or largely molten solution (i.e., magma) that solidifies deep within the crust to form an intrusive rock, or is transported to the surface prior to completely solidifying to form an extrusive rock. Intrusive rocks cool slowly at depth to form relatively coarse-grained bod-

ies such as granite, granodiorite, and gabbro. Extrusive rocks are rapidly quenched, producing glassy or fine-grained volcanic ash and lava flows, such as rhyolite, rhyodacite, and basalt. Granites-rhyolites consist mainly of quartz and alkali feldspar, whereas gabbros-basalts contain olivine, pyroxene, and Ca-rich plagioclase. Rocks with mineralogical and bulk chemical bulk compositions intermediate between these end members are common.

2. *Sedimentary*—Such deposits form by the mechanical settling of particulate matter or precipitation of a solute from a fluid, typically water. Progressively finer grained clast sizes make up conglomerates, sandstones, siltstones, and mudstones (i.e., shales); the coarser sedimentary clasts are rich in quartz and feldspars, whereas the finer grained mudstones are dominated by clay minerals. Many chemical precipitates, but not all, are biologically generated; the major chemical sedimentary rocks are limestones (calcite) and chert (quartz).

3. *Metamorphic*—This group of rocks has been transformed at depth in the crust by deformation and/or physicochemical conditions that were distinctly different from those attending the formation of the preexisting igneous and sedimentary (or metamorphic) rock types. Greenstone, serpentinite, marble, gneiss, and slate are familiar examples of recrystallized (metamorphosed) basalt, mantle lithosphere, limestone, granite, and shale, respectively.

Surface interactions of minerals, mineraloids, and rocks with agents of the biosphere, atmosphere, and/or hydrosphere result in alteration of chemically reactive earth materials to produce a thin veneer of clay-rich soil. The process is termed weathering and results in removal and transportation of earth materials as soluble species in aqueous solution and as insoluble particles entrained in moving fluids (wind, water, and ice). The residue left behind over time builds up a soil profile. It is such weathering products that in many cases provide the ready supply of both nutrients and toxic chemical species that influence the existence of life in general and human health in particular. Geological mapping and remote sensing techniques provide the enhanced spatial understanding of the areal disposition and concentration of surficial earth materials that are an essential component of epidemiological investigations of environmentally related diseases and human senescence.

Soil and the Vadose Zone

The human environment is heavily dependent on the continuum between soil, water, and air that is located at the earth's surface. Ultimately this continuum—and the interactions between the physical, chemical, and biological properties of each component—moderates many of our activities. The geological zone between the land surface and subsurface groundwater—the *vadose zone*—consists of unsaturated organic and earth materials. A subset of this vadose zone is the near-surface soil environment, which is in direct contact with both surface water and the atmosphere.

Soil directly and indirectly influences our quality of life—it is taken for granted by most people but is essential for our daily existence. It is responsible for plant growth and for the cycling of nutrients through microbial transformations, and has a major effect on the oxygen/carbon dioxide balance of the atmosphere. Because of our reliance on soil, any disturbance of soil or the vadose zone, or modification of natural soil-forming processes, has the potential for adverse public health effects. Soil also plays a critical public health role in regard to pollutants that have been disposed of at the earth's surface, as they can promote or restrict transport to groundwater, the atmosphere, or food crops.

Soil is a complex mixture of weathered rock particles, organic residues, air, water, and billions of living organisms that are the end product of the interaction of the *parent rock material* with *climate, living organisms, topography,* and *time*—the five soil-forming factors. The soil layer can be as thin as a few inches or may be hundreds of feet thick. Because soils are derived from unique sources of parent material under specific environmental conditions, no two soils are exactly alike—there are thousands of different kinds of soils within the United States.

Soils can be acidic (pH <5.5), neutral (pH of 6–8), or alkaline (pH >8.5). Soil pH affects the solubility of chemicals in soils by influencing the degree of ionization of compounds and their subsequent overall charge. The extent of ionization is a function of the pH of the environment and the dissociation constant (pK) of the compound. Consequently, soil pH can be critical for affecting the transport of potential pollutants through the soil and vadose zone and can also affect the transport of viruses with different overall charge. In high rainfall areas, the combination of acidic components and residues of organic matter, together with the leaching action of percolating water, leads to acidic soils. Conversely, soils in arid areas are more likely to be alkaline because of reduced leaching, lower organic contents, and the evaporative accumulation of salts.

Soil normally consists of about 95–99% inorganic and 1–5% organic matter. The inorganic material is composed of three particle size types—sand (0.05–2 mm), silt (0.002–0.05 mm), and clay (<0.002 mm; i.e., <2 microns)—that result from the weathering characteristics of the parent rock. In some geological terrains (e.g., some igneous and glacial areas), soils also contain larger (>2 mm) gravel- and cobble-sized particles inherited from the parent rock type. The percentage of sand, silt, and clay in a particular soil determines its texture (see Figure 2.3), which affects many of

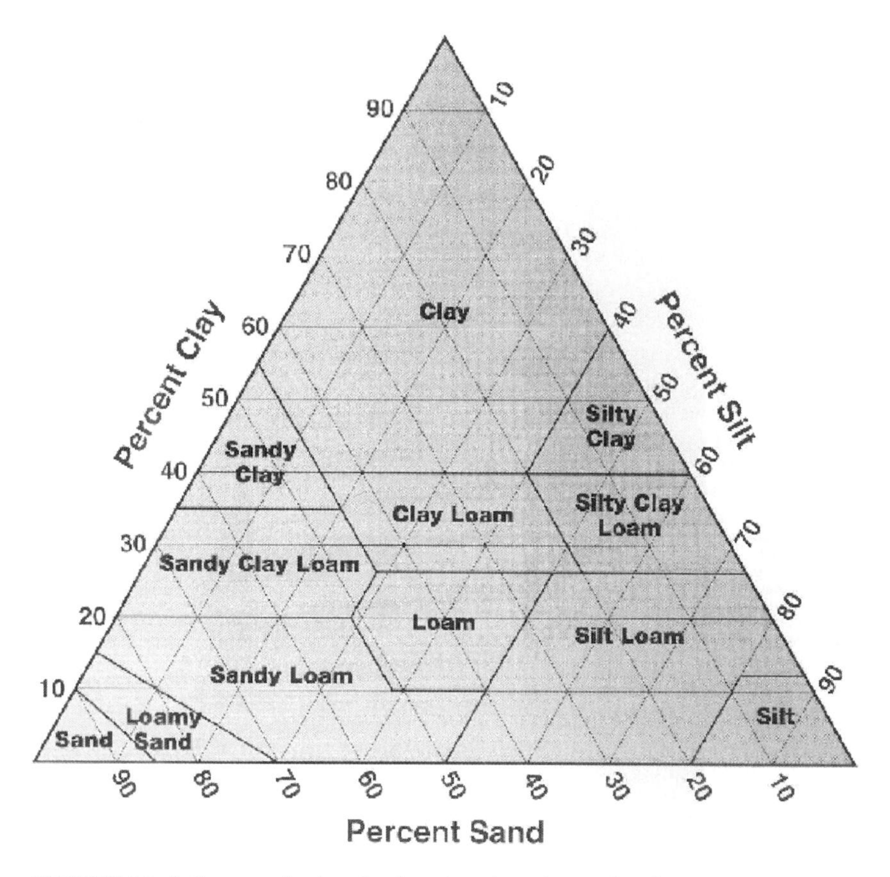

FIGURE 2.3 Soil textural triangle showing the relationship between soil texture categories and particle size. These textural classes characterize soil with respect to many of their physical properties.
SOURCE: Pepper et al. (2006).

the physical and chemical properties of the soil. Of the three primary particle types, clay is by far the dominant component for determining a soil's properties because of the greater number of clay particles per unit weight. The increased surface area of soils with higher clay concentrations leads to increased chemical reactivity of the soil. In addition, clay particles are the primary soil particles that have an associated electric charge. This is the basis for a soils cation-exchange capacity (CEC), which is normally a negative charge that occurs because of isomorphic substitution or ionization of hydroxyl groups at the edge of the clay lattice.

Differences in the partitioning of elements among the different par-

TABLE 2.1 Size Fractionation of Soil Constituents

Mineral Constituent	Size	Organic and Biological Constituents	Surface Area
Sand Primary minerals: quartz, silicates, carbonates	2 mm	Organic debris	0.0003 m^2 g^{-1}
Silt Primary minerals: quartz, silicates, carbonates	50 μm	Organic debris and large microorganisms: fungi, actinomycetes, bacterial colonies	0.12 m^2 g^{-1}
Granulometric Clay Microcrystals of primary minerals Phyllosilicates: inherited (illite, mica) transformed (vermiculite, high-charge smectite) neoformed (kaolinite, smectite) Oxides and hydroxides	2 μm	Amorphous organic matter: humic substances, biopolymers	30 m^2 g^{-1}
Fine Clay Swelling clay minerals Interstratified clay minerals Low range order crystalline compounds	0.2 μm	Small viruses	3 m^2 g^{-1}

NOTE: Data in Surface Area column represent specific surface area using a cubic model.
SOURCE: Modified from Pepper et al. (2006).

ticle size classes is an important component of understanding potential health effects from soils. Elemental variations result both from the mineralogical and geochemical characteristics inherited from parent rock materials and, particularly for the clay fraction, from macro and trace elements introduced by contamination. The distribution of inorganic and organic constituents among the different soil particle classes is summarized in Table 2.1.

The three types of primary particles do not normally remain as individual entities. Rather, they aggregate to form secondary structures, which occur because microbial gums, polysaccharides, and other microbial metabolites bind the primary particles together. In addition, particles can be held together physically by fungal hyphae and plant roots. These secondary aggregates, which are known as "peds," can be of different sizes and shapes, and give the soil its structure. Pore space within the aggregate structure (intraggregate pore space) and between the aggregates (inter-aggregate pore space) is crucial to the overall soil architecture. Pore space also regulates water movement and retention as well as air diffusion and microsite redox potentials.

Organic Matter in Soils

Organic compounds are incorporated into soil at the surface via plant residues such as leaves or grassy material. These organic residues are degraded by soil microorganisms, which use the organic compounds as food or microbial substrate. The main plant constituents—cellulose, hemicellulose, lignin, protein and nucleic acids, and soluble substances such as sugars—vary in their degree of complexity and ease of breakdown by microbes. In general, soluble constituents are easily metabolized and break down rapidly, whereas lignin, for example, is very resistant to microbial decomposition. The net result of microbial decomposition is the release of nutrients for microbial or plant metabolism, as well as the particle breakdown of complex plant residues.

The nutrient release that occurs as plant residues degrade has several effects on soil. The enhanced microbial activity causes an increase in soil structure, which affects most of the physical properties of soil, such as aeration and infiltration. The stable humic substances contain many constituents that contribute to the pH-dependent CEC of the soil. In addition, many of the humic and nonhumic substances can complex or chelate heavy metals and sorb organic contaminants. This retention affects their availability to plants and soil microbes as well as their potential for transport into the subsurface. Overall, the physical constituents control many of the chemical reactions that occur within soil, with fine particles (~2 μm)—inorganic granulometric and fine clays as well as organic matter that results from microbial decomposition of plant residues—being particularly important.

Conversely, it is the soil microflora that control biochemical transformations in soil. Interestingly, the organic and biological constituents of soil mimic the mineral constituents with respect to size (Table 2.1). In order of increasing size, the major soil biota consists of viruses, bacteria, fungi, algae, and protozoa. As size decreases, the number of organisms increases to staggeringly large numbers—a gram of soil literally contains billions of organisms. The physical heterogeneity of soil results in microenvironments that allow diverse microbial communities to coexist in close proximity. Overall, the variable terminal electron requirements of aerobic and anaerobic microbes coupled to variable nutritional requirements (autotrophy and heterotrophy) result in extraordinary soil biological diversity. The soil microflora are responsible for many of the biochemical processes essential for human life, including plant growth, products for human health, and groundwater protection (Stirzaker et al., 1996; Bejat et al., 2000; Strobel and Daisy, 2003). Conversely, soils also contain human pathogens and are a source of bacterial antibiotic resistance.

Gases and Liquids in Soils

Because soil and the atmosphere are in direct contact, most of the gases found in the atmosphere are also found in the air phase within the soil (the soil atmosphere)—oxygen, carbon dioxide, nitrogen, and other volatile compounds such as hydrogen sulfide or ethylene. The concentrations of oxygen and carbon dioxide in the soil atmosphere are normally different from those in the atmosphere, reflecting oxygen use by aerobic soil organisms and subsequent release of carbon dioxide. In addition, gaseous concentrations in soil are altered by diffusion of oxygen into soil and carbon dioxide from soil. Because microbial degradation of many organic compounds in soil is carried out by aerobic organisms, the presence of oxygen in soil is necessary for such decomposition. Oxygen occurs either dissolved in the soil aqueous solution or in the soil atmosphere.

The total amount of pore space depends on soil texture and structure. Soils high in clays have greater total pore space but smaller pore sizes. In contrast, sandy soils have larger pore sizes, allowing more rapid water and air movement. Aerobic soil microbes require both water and oxygen, which are both found within the pore space, and therefore soil moisture content controls the amount of available oxygen in a soil. In soils saturated with water, all pores are full of water and the oxygen content is very low. In dry soils, all pores are essentially full of air, so the soil moisture content is very low. In soil with moderate moisture content, both air (oxygen) and moisture are readily available to soil microbes. In such situations, soil respiration via aerobic microbial metabolism is normally at a maximum.

Because they are unsaturated, vadose zones generally are aerobic. However, due to the heterogeneous nature of the subsurface, anaerobic zones can occur in clay lenses and so both aerobic and anaerobic microbial processes may occur in close proximity. At contaminated sites, volatile organic compounds can be present in the gaseous phase of the vadose zone. For example, chlorinated solvents, which are ubiquitous organic contaminants, are volatile and are typically found in the vadose-zone gaseous phase below hazardous waste sites.

Soil as a Genetic Resource

Soil is the home for billions of microorganisms with unimaginable diversity. It has been estimated that 4,000 completely different genomes of standard bacteria (Torsvik et al., 1990) and at least 1 million different species of fungi (Gunatilaka, 2006) are present in soil. Of these organisms, less than 1% of bacterial species and 5% of fungal species have been identified (Young, 1997). Despite this limitation, soil has proven to be a treasure chest of natural products critical to maintaining human health and welfare. Of the soil microbes, actinomycetes and fungi have proved to be particularly rich sources of metabolites with novel biological activities. The fungal antibiotic penicillin and the actinomycete antibiotic streptomycin were the first antibiotics to be discovered from soil. More recently, interest has centered on rhizosphere bacteria and endophytic microbes as a new source of natural products including antibiotics.

Biogeochemical Cycling

Over the past century, industrialization has accelerated and the environmental consequences are ubiquitous. Although some believe that environmental changes on our planet are caused entirely by anthropogenic excesses, the earth's environment has in fact undergone profound changes—caused by processes such as those briefly described above—throughout its history. The superposition of earth materials—the stratigraphy of sediments and rocks—provides a record of past physical and biological processes. These in turn define the directions and magnitudes of the physical and biological changes that provide the context for modern conditions. This section notes the diversity of biological systems, emphasizing the metabolic processes of microorganisms and their roles in consuming earth materials to produce metabolites and influence the environment.

Diverse and abundant microorganisms, such as bacteria, viruses, fungi, algae, protozoa, and other groups of single-celled organisms, affect nearly all aspects of the environment from the atmosphere to deep within sediments beneath the ocean floor. To many people, microorganisms are viewed as harbingers of disease. But food technologists know that the life processes of bacteria and yeast are essential for the production of many dairy products, baked goods, and alcoholic beverages. Soil scientists study the complex symbioses involving microbes and plants. Few of us realize the importance of microorganisms in the formation of many natural resources, including petroleum, carbonate and silica minerals, and metallic ore deposits. They play crucial roles in the global cycling of carbon, sulfur, and nitrogen and in maintaining the composition and the dynamic equilibrium among the chemical species present in the oceans, soils and rocks, and the atmosphere.

Six elements are of crucial importance for all life on earth—carbon (C), hydrogen (H_2), oxygen (O_2), nitrogen (N_2), sulfur (S), and phosphorus (P). These elements, collectively referred to as CHONSP, are essential components of the building blocks of all cells from unicellular bacteria to multicellular mammals. In different combinations and ratios, they are present in carbohydrates, lipids, proteins, and nucleic acids. They combine to form skeletal materials such as lignin and cellulose in plants, chitin in insects and crustaceans, and keratin in mammals and—with the calcium cation (Ca)—form apatite in bones and calcite in invertebrates. A wide spectrum of metabolic systems has evolved to efficiently recycle these elements. Accordingly, the products of one set of biochemical processes are used as reactants for another set, thus ensuring that the elements are not irreversibly bound in a form that is unavailable to living matter.

Biogeochemical cycling is the recycling of elements by organisms in the context of geological processes. Various microorganisms participate in every chemical transformation of CHONSP. These tiny, incredibly abundant and diverse organisms are the workhorses of biogeochemistry. They degrade previously synthesized organic material or form new organic substances by fixing carbon dioxide, both photosynthetically and in the absence of light. In carrying out their wide range of metabolic functions, they consume and then release each element, thus returning it to the biosphere, the hydrosphere, the atmosphere, and/or the solid earth. An instructive example is represented by the development of soil—the substrate on which the terrestrial food chain is based.

HUMAN PHYSIOLOGICAL PROCESSES

All of us share a bipedal upright structure and a body with distinctive anatomical parts that have discrete morphologies and functions. Acting in concert, these components maintain a state of balance (homeostasis) within the organism. The loss of body structural integrity and function can occur when homeostasis is perturbed by internal factors or by physiological response to hazardous materials in the environment. Most earth materials—solids, liquids, or gases—are essential for the body or are benign. A few can become harmful, especially if in elevated amounts, where they impinge or enter the body and disturb the normal functions of the organs. There are three usual routes of exposure to earth materials—respiration (through the nose or mouth and into the breathing apparatus), ingestion (through the mouth into the digestive system), and dermal (through the skin).

The circulatory system that transports blood to all organs and tissues is the pathway for disseminating the nutrients—and any hazardous substances—throughout the body. Most hazardous substances that enter the body are either excreted, or protective mechanisms present in our bodies that refresh the cell populations in tissues will act over time to diminish the hazard. Unfortunately, some normal protective processes may exacerbate the action of the hazard, aiding, for example, local accumulation of the hazardous species, or provide unwanted long-term consequences such as occurrences of scarring from deposition of particulate matter in the lung. Although the level and duration of exposure (see Box 2.1) that can lead to disease are unique to an individual, there are some hazards—such as asbestos and arsenic—that are known public health issues at minimal exposure levels.

Human anatomy is the starting point within which complicated tissue, cellular, and molecular-level reactions take place that define specific diseases. Details of these specific reactions and diseases are supplied by specialists (e.g., Warwick and Williams, 1973; Goldman and Ausiello, 2004; Cohen, 2005) and are beyond the scope of this report. However, partial or total inhibition of one of these body systems would likely have dire results that we can all appreciate. A simplified overview of the physiology of three body systems and their normal functions—the respiratory, digestive, and integumentary (skin) systems—provides the human context for this report.

Inhalation Pathways

The human respiratory system consists of organs and tissues in the upper trunk that permit us to breathe ambient air, extract oxygen, and respire carbon dioxide back into the atmosphere. Under both voluntary and involuntary nervous system and muscular control, the continual motions of inspiration and exhalation of air provide the oxygen essential for metabolic activity throughout the body. The passageways from the nose or mouth through to the lungs, with their specialized air sacs or alveoli, are illustrated in Figure 2.4. These tiny balloonlike structures are surrounded by arterioles and venules where the essential gas exchange—into and from the blood—actually takes place. Hazardous materials in ambient air may access the inner portions of the body through the respiratory system sampled with each breath.

The respiratory system has a series of built-in defenses, natural mechanisms, and strategies that are very effective for minimizing inadvertent transport of a wide range of potential hazards, whether they are solids, liquids, or gases. The first line of defense is the nose, where hairs, and/or a sneeze, will expel unwanted materials. Ciliate cells that popu-

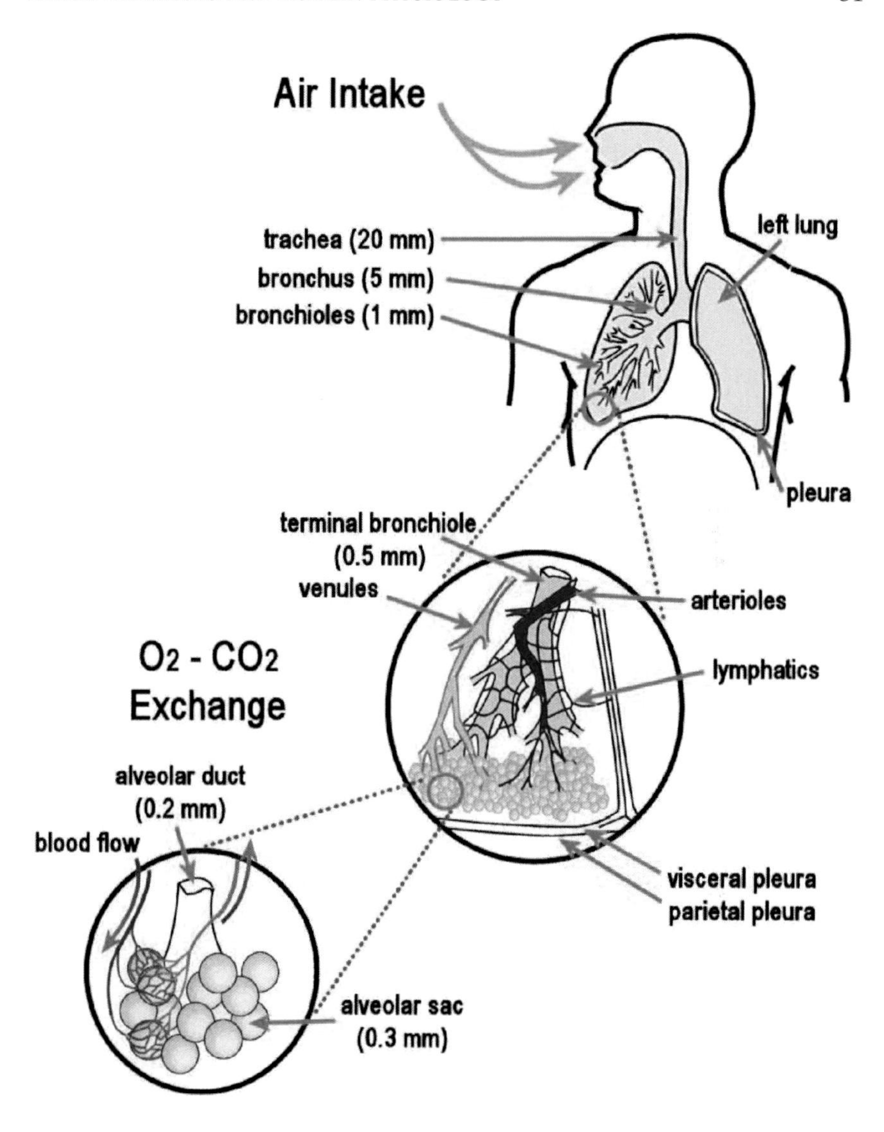

FIGURE 2.4 Schematic diagram of the human respiratory system, showing the gross anatomy of the lung, the covering membranes (pleura), airways, and air sacs (alveoli). The average diameter of portions of the air flow system are indicated—trachea, 20 mm; bronchus, 8 mm; terminal and respiratory bronchioles, 0.5 mm; alveolar duct, 0.2 mm; and alveolar sacs, 0.3 mm.
SOURCE: Modified after Skinner et al. (1988).

BOX 2.1
Exposure Assessment in Epidemiological Studies

Epidemiological studies of environmental exposure—studies of adverse health outcomes in human populations—are designed to quantify the risk of disease at particular levels of exposure to the environmental threat. The success of such studies to detect an effect, or conversely to demonstrate that health consequences of specific exposures are likely to be minimal, depends on both the accuracy of the health outcome diagnosis and the accuracy of estimates of exposure to the environmental insult.

Depending on the health outcome under study, there can be a wide range of diagnostic accuracy. However, in countries with advanced systems of medical care, diagnostic accuracy is typically quite high for health effects such as cancer, major birth defects, myocardial infarction (heart attack), and many neurological diseases, and the inaccuracies that do occur typically do not limit the ability to accurately quantify the effects of environmental exposures. Far more important in influencing the success of such studies is inaccuracy in exposure assessment, usually referred to as "misclassification of exposure." That is, persons with high exposures can erroneously be classified as having had low or no exposure and persons with low exposure as having had high exposure. In addition, many studies use indicators of exposure rather than actual measurements of exposure, and such indicators can be less precise. Even in instances of equivalent exposure, differences in individual physiological characteristics (e.g., respiration rate, the capacity to metabolize specific xenobiotic compounds) can cause differences in effective dose that can produce varying health outcomes.

How does such misclassification affect the findings of an epidemiological study? The basic measure of effect used by most epidemiological studies is the relative risk—an estimate of the ratio of disease (or death) rate among exposed persons relative to the unexposed rate. For example, the relative risk for lung cancer among pack-a-day lifetime smokers is about 15, meaning that smokers have about 15 times the rate of lung cancer as people who have never smoked. There is a gradient of risk in most real-world situations, with the relative risk increasing with the level of exposure. When exposure misclassification occurs in epidemiological studies without regard to whether subjects are diseased or not diseased, the result is usually to lower the estimates of the effect. Given a "true" relative risk, it is possible to calculate what a researcher would actually observe in an epidemiological study, given various levels of exposure misclassification. In a simplified example, where subjects are either "exposed" or "unexposed" (i.e., no gradient of exposure), and the correlation coefficient is 0.60 between the "true" and estimated exposures, a relative risk of 2.0 would appear to have a relative risk of 1.6 (Vineis, 2004). In some situations, such a difference could determine whether or not an exposure is identified as a significant risk factor. Thus, misclassification of exposure in epidemiological studies can have a major detrimental effect, and it is highly worthwhile to place major resources into the exposure assessment efforts when investigating environmental or occupational exposures.

Earth scientists frequently play a crucial role in providing measures or estimates of exposure in epidemiological studies of environmental contaminants. A good example is the collaboration between the National Cancer Institute (NCI) and the U.S. Geological Survey (USGS) in an epidemiological study of bladder cancer in northern New England, which also includes collaborators from Dartmouth Medical School and the state health departments of Vermont, New Hampshire, and Maine. Since 1950, bladder cancer mortality in New England for both males and females has been above the U.S. average. Because some areas of New England have modestly elevated levels of arsenic in groundwater, arsenic in drinking water has been suggested as a reason for these excess bladder cancer rates. The identification of arsenic in drinking water as a cause of bladder cancer in parts of Taiwan, Chile, and Argentina, where it occurs (or has occurred) at considerably higher levels, has been proposed as supporting evidence for this suggestion. The NCI recently completed fieldwork in New England for a population-based case control study, over a three-year period, which involved about 1,200 newly diagnosed cases of bladder cancer and 1,200 disease-free persons. A water sample was collected at the current homes of patients and control subjects, and information about past residences and their water sources, together with much additional information, was obtained in personal home interviews. The USGS located past wells used by the subjects and, wherever possible, collected water samples. USGS scientists also developed geologically based statistical models to estimate arsenic levels in well water where sample collection was not feasible (Ayotte et al., 2006).

late the surface of the upper bronchi constantly beat upward, also acting to remove substances that may not have been propelled out by a sneeze. They are often swallowed or expectorated with continually locally supplied mucus. If a particle is small enough to bypass these expulsion efforts, it may become lodged in the tissues along the passageways or in the lung tissue, where special cells such as macrophages may engulf the intruding particle and transport it through the lymph system to the lymph nodes. Another protection option occurs when offending materials become isolated by being engulfed in fibrous collagenous protein (collagen) produced by cells known as fibroblasts. This mechanism is part of the normal repair strategies employed by biological systems subjected to trauma (Skinner et al., 1988).

The structures of the respiratory system from the macro level (i.e., the muscles and the connective tissues forming the pleura that enclose and define the lung cavity), to the specialized cells for transport of gases and those that act in natural defense of the system, have a multitude of normal biological reactions. Some may become pathological and disadvantageous, decreasing the availability of oxygen. Lung disease is a product of the sensitivities and lung physiology of an individual and the duration and exposure level to a hazardous substance.

There are a host of potentially harmful substances whose presence can produce breathlessness, hacking cough, or difficulties in breathing. Respiratory malfunctions are usually diagnosed through physical examination, including auscultation (listening with a stethoscope), radiological examination (X-ray imaging) of the chest, and patient exposure histories. Clinical evaluation may be able to identify likely potential offenders through the specificity of organ responses, which may in turn be confirmed by histological analysis. With repeated exposures, especially at high doses over a long time, the incorporation of foreign materials usually causes cells in the respiratory system to die or to mutate, and diseases such as chronic obstructive pulmonary disease (COPD) or cancer may result. Smoking, of course, is the most egregious example of deliberate inhalation of hazardous material.

Inhalation of gaseous, liquid, or particulate matter can alter and perhaps reduce pulmonary function, especially if this foreign material causes irritation in the airways because normal clearance mechanisms are compromised (e.g., by smoking). Protection from irritation of the delicate tissues is part of the normal breathing apparatus—cilia in the nose and upper respiratory tract, coughing and expectoration, cellular (e.g., macrophage) activities all aid in the expulsion or isolation of potentially damaging foreign material. If these multiple preventive mechanisms prove unable to expel or mitigate the adverse action of the foreign materials on the respiratory system, chronic coughs, bronchitis, or asthma may result from such irritations and/or continued exposures (Skinner et al., 1988; Koenig, 1999). Particle size determines whether particles are respirable (<10 μm) or inhalable[1] (10–100 μm). Respirable particles tend to be deposited at the junctions of the respiratory bronchioles in humans (see Figure 2.4). The airways and the alveolar regions, where the air/carbon dioxide exchange takes place, are affected in diseases such as asbestosis and silicosis. Studies have estimated that with each 10 μg m^{-3} increase in PM_{10} above a base level of 20 mg m^{-3}, daily respiratory mortality increases by 3.4%, cardiac mortality increases 1.4%, hospitalizations increase 0.8%, and school absenteeism increases 4.1% (Vedel, 1995).[2] Particulate matter can also exacerbate the production of photochemical smog, the end product of the interaction of smoke, volatile hydrocarbons, and fog, under the influence of ultraviolet light.

Ingestion Pathways

The essential activities of the gastrointestinal system are absorption of nutrients and elimination of waste. Whenever something is ingested, at least a portion is usually digested and absorbed and is distributed to all parts of the body by the circulation system, either in its original form or as modified by chemical transformation in the liver or elsewhere.

Mastication of solid food by the teeth reduces the size so that it can be swallowed and enter the gastrointestinal tract. Saliva and other gland secretions that empty into the mouth moisten the chewed food, providing enzymes like amylase that dissolve starches and add to the disintegration. Complete solubilization of all the large molecules in foodstuffs leading to release of soluble constituents (e.g., small ions, or molecules) requires biochemical activity that commences in the saliva and continues into the small intestine. Specialized gland cells that line the inside of the tract secrete chemicals that break down the solids, while others are involved in absorption and transmission of the "chemicals," some of which are nutrients, and make them available to the circulatory system. The mobilization of ingested materials, and the extraction and adsorption of nutrients, is the focus of the digestive tract. The tract is effectively a flexible tube that ranges in length up to about 9 m from the mouth to the anus in an adult.

The organs in the digestive tract are supported by fibrous connective tissue, and in some parts of the system there are double layers that also

---

[1]Inhaled particulate material enters the nose, throat, and upper respiratory tract, whereas respirable particulate material is able to penetrate deep into the lungs.

[2]PM_{10} denotes particulate matter smaller than 10 μm in effective diameter.

carry blood vessels, lymph vessels, and nerves. Each of these organs performs specialized activities with specialized cells and each produces proteins and enzymes that have minor and trace elements as cofactors. Ingestion of hazardous materials or elements in food or water may induce local aberrations in the normal operations of each of these essential organs. High concentrations of arsenic (see Chapters 4 and 5), for example, must pass through the gastrointestinal system and can influence local biochemical reactions.

Clinically, examination of blood and urine is traditionally used to assay the amount of a potential hazard that has been ingested. However, despite the fact that fluids are consistently exchanged throughout body compartments, if there is a potential for accumulation intracellularly or within certain tissues, then these trace levels are not easily assessed. The fate of absorbed nutrients or hazardous materials may be modified in the body by the liver and dumped into the intestine for elimination in the feces. Alternatively they may be filtered and recycled by the kidney or actively secreted in the urine. Some unwanted chemicals are known to be stored in fats (e.g., DDT) or other cells in relatively small amounts.

Skin Absorption Pathways

The skin is the largest organ of the human body. More properly known as the integumentary system, the skin is composed of thin layers of tissue that cover the entire body and in some instances presents early clues to an individual's health status. The outermost area of skin—the epidermis—contains keratinocytes[3] that produce stratified squamous epithelium layers and a few other types of cells. The base of the epidermis is interdigitated with the underlying dermis, mostly connective tissue of mesodermal origin. The dermis supplies some nourishment to the lowest layers of the epidermis and is a framework of connective tissue that contains blood vessels, nerve endings, glands, and the roots of hair follicles. The skin has variable thickness—it may reach 1.5 mm at the soles of the feet and palms of the hands, whereas in other areas (such as around the eyes) the skin is a thin (< 0.1 mm) delicate covering that is relatively elastic.

Although one function of the skin is to assist in the formation of vitamin D, the major contribution of skin to human health is to protect the body from invasions of pathogens, prevent dehydration, and keep body temperature constant. The optimal performance of this organ requires continual regeneration to maintain complete body coverage to minimize

[3]Keratinocytes are the major cell type in the epidermis, and the different epidermal layers are distinguished on the basis of keratinocyte morphology.

adventitious fluid and water loss or infection. Dermal toxicity results from local tissue responses through direct contact of a substance with skin or, alternatively, may represent a manifestation of systemic toxicity following ingestion or inhalation. Allergic contact dermatitis induced by nickel is an example of a local tissue response (Centeno, 2000).

GEOAVAILABILITY, BIOAVAILABILITY, AND BIOACCUMULATION

Many chemical elements occur in living tissues in such small concentrations that they are referred to as trace elements. Some trace elements are essential for human life because of their role as catalysts in cellular functions involving metabolic or biochemical processes. At present, less than one-third of the 90 naturally occurring elements obtained from the air, water, and food are known to be essential to life. The mineral elements currently considered essential for human health and metabolism include the major ions/anions sodium (Na^+), calcium (Ca^{2+}), chlorine (Cl^-), magnesium (Mg^{2+}), potassium (K^+), silicon (Si^{4+}), sulfate (SO_4^-), and nitrate (NO_3^-); trace elements such as phosphorus (P), iodine (I), and fluorine (F); and metals/metalloids such as iron (Fe), zinc (Zn), copper (Cu), manganese (Mn), vanadium (V), selenium (Se), cobalt (Co), nickel (Ni), chromium (Cr), tin (Sn), and molybdenum (Mo) (Moynahan, 1979). Some of these species occur predominantly in silicate minerals (e.g., Mn, boron [B]), some in silicates and sulfides (e.g., Zn, Se) or as trace element impurities in phyllosilicate minerals, and some predominantly as sulfides (e.g., Cu, Mo); others (e.g., Fe) are ubiquitous (Combs, 2005). The bioavailability and bioassimilation of these essential mineral elements are dependent upon each of their unique physiochemical properties.

Various definitions have been used to describe trace element concentrations within earth science materials and humans (see Box 2.2). Interestingly, both earth scientists and public health professionals have distinguished total elemental concentrations from bioavailable concentrations. In essence, bioavailable concentration in soils refers to the concentration of an element in solution that can be taken up by plants or microorganisms. Within the human body, the term "bioavailable" refers to the amount of a particular element that can be absorbed by the body and influence human health and welfare. For earth science materials and the human body, bioavailable concentrations are always less than total elemental concentrations.

Many trace elements are metabolic requirements for humans, so too small an amount of these elements will result in deficiency (see Figure 2.5A). Although the biological response is optimal at higher concentra-

BOX 2.2
Bioavailability Definitions

Earth Science Definitions

Bioavailability—that fraction of an element or compound in solution that can be taken up by plants or soil micro-organisms.

Biomagnification—the bioaccumulation of a substance up the food chain by transfer of residues of the substance in smaller organisms that are food for larger organisms in the chain.

Geoavailability—that portion of an element or compound's total content in an earth material that can be liberated to the surficial or near-surface environment (or biosphere) through mechanical, chemical, or biological processes.

Health Science Definitions

Bioaccessible fraction—that fraction of a metal or a metal compound that is soluble in various body fluids (gastrointestinal, respiratory, perspiration, etc.). Solubility is dependent on individual physiology.

Bioavailable fraction—that fraction of a metal or metal compound that is absorbed by the body and transported within the body to a site of toxicological action. Absorption and transport are both dependent on individual physiology.

tions, at still higher concentrations the element may become toxic. Toxicologists express the level at which these phenomena are observed in several different ways. The first is the No Observable Adverse Effect Level (NOAEL) or No Observable Adverse Effect Concentration (NOAEC). At higher concentrations, the biological response is expressed as the Lowest Observable Adverse Effect Level (LOAEL) or Lowest Observable Adverse Effect Concentration (LOAEC). The dose-response curve lacks a deficiency zone for nonessential elements (see Figure 2.5B).

Element toxicity depends on the bioavailability of the element, its distribution in the body, the physical and chemical form of the element, and its storage and excretion parameters. In recent years, considerable interest has been focused on assessing the human health risk posed by metals, metalloids, and trace elements in the environment. It has long been recognized that large areas of the globe contain human populations characterized by having trace element excess, deficiency, or chronic poisoning (e.g., Selinus et al., 2005).

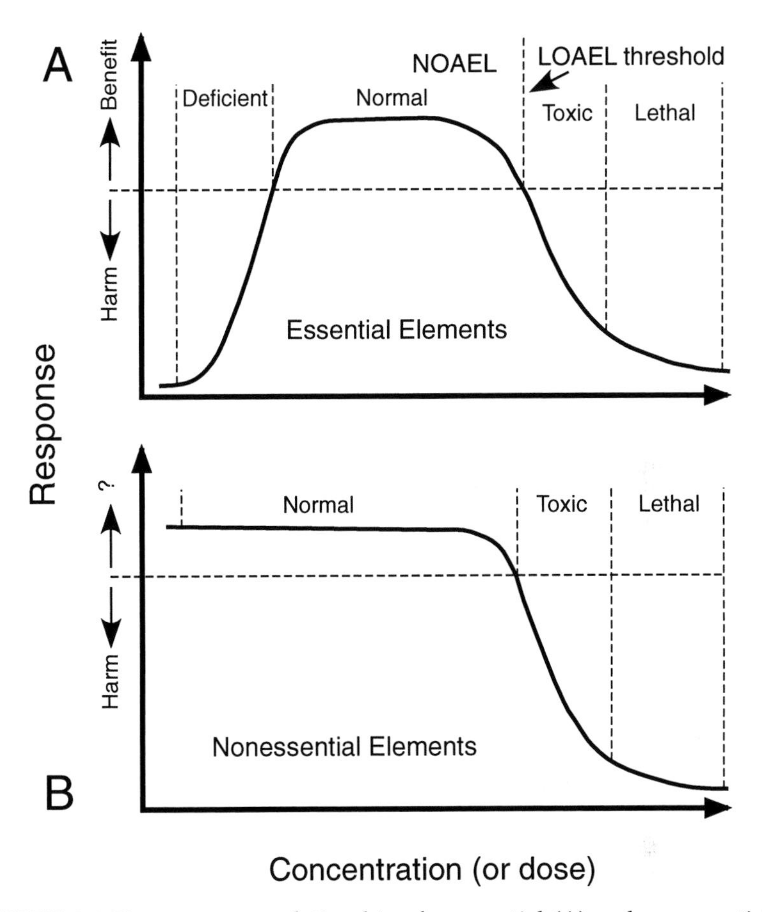

FIGURE 2.5 Dose-response relationships for essential (A) and nonessential (B) elements.
SOURCE: Modified after Adriano (2001).

Section II

Exposure Pathways

3

What We Breathe

The air we breathe is a heterogeneous mixture, a composite of gases, airborne solids, and liquids. Aerosols (mixtures of liquids and gases, or liquids and various chemical compounds including solids) and particulate matter are present in the air in concentrations that are variable over time and space. This report focuses on airborne natural materials derived from earth sources—the natural mineral, gaseous, and biological constituents of the earth's surface that occur in the atmosphere and can cause either beneficial or adverse effects to human health and welfare.

Natural contaminants, such as wind-blown dust from arid areas, can carry bacteria and fungi. Such complexes of inorganic and biological particulate matter can travel within the troposphere for long distances (see Figure 3.1), even around the globe, in relatively short times. Other intermittent natural sources of harmful aerosols with obvious and immediate health impacts are emissions from volcanoes, including particulate matter as well as volcanogenic gases such as sulfur dioxide (SO_2) and carbon dioxide (CO_2). Active volcanoes, with their associated vents and fumaroles, have a long record of affecting populations worldwide; for example, the Icelandic eruption of Laki in 1783–1784 (see Box 3.1) caused many deaths in Europe, particularly of the infirm and the young (Grattan et al., 2005).

This chapter focuses on both *direct* health effects, such as the inhalation of suspended particulate matter (rock and soil particles) or gases (volcanic and biogenic gases, radon) that pose a health benefit or risk, and *indirect* effects, such as the inhalation of bacteria attached to soil particles.

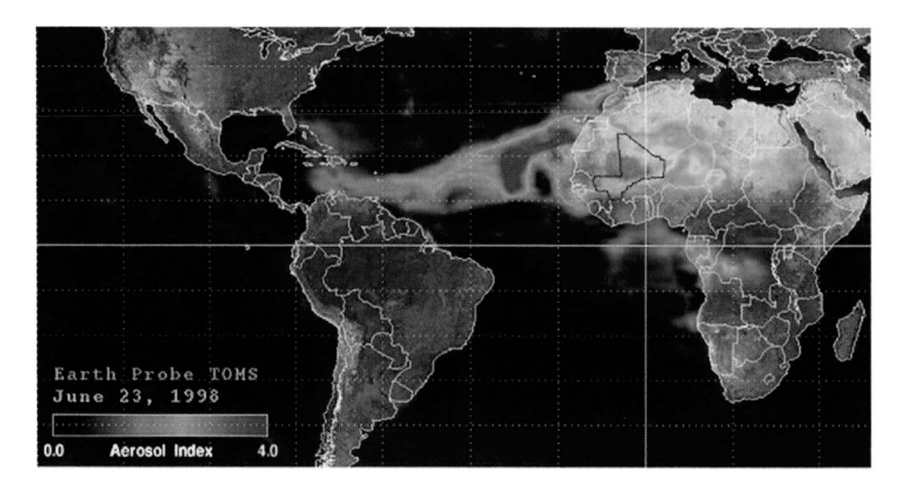

FIGURE 3.1 Satellite image, acquired by NASA's Earth-Probe TOMS (Total Ozone Mapping Spectrometer) satellite instrument on June 23, 1998, showing an African dust event extending from the western Sahara to the Caribbean and Florida. The TOMS aerosol index is a relative measure of absorbing aerosol particles suspended in the atmosphere; the higher the index (warmer colors), the greater the particle load.
SOURCE: Kellogg et al. (2004).

INHALATION OF PARTICULATE MATERIAL

Only a few natural materials are inherently hazardous, and few are sufficiently accessible or mobile to pose a health risk in unperturbed landscapes. Asthma—a chronic condition consisting of airway inflammation and bronchoconstriction—is an example where earth material particulate matter can have adverse health effects. Current research suggests that asthma is caused by a combination of genetic and environmental factors but that particulate matter inhalation increases the severity of asthma (NASA, 2001).

Humans have a history of using specific elements from the environment, thereby changing the natural surficial distributions of rock material, soil, and botanical ground cover. At many mining sites, for example, this has led to several-fold increases of earth-sourced airborne particulate matter and increased human exposure to potentially hazardous materials. In such environments, ground-based ambient air sample data combined with spatially located health data can demonstrate the impacts of such exposure. Surface measurements of aerosol emissions identify the source and site of origin and, through analyses of the particulate matter, the composition of the potential hazards. On a local scale, integration of health reactions with surficial sample analyses from multiple sites provides an early warning system. Satellites are increasingly able to detect the initiation of potential hazardous aerosols, and with increased sensitivities these remote sensing data have the potential to be combined with earth process models to provide a global warning system (Torres et al., 1998, 2002).

The reactions of individuals to air pollutants are variable. Low-level concentrations of airborne particulate matter and chemicals may require decades of exposure before the adverse effects are even noticed. In contrast, even brief exposure to an airborne pathogen can result in immediate illness. Although the problem of impure air has beset most societies throughout recorded history, including the ancient Greeks and Romans, it was not until the latter half of the twentieth century that many countries took action to address the problem (Matthias, 2005). For example, in the United States, Congress in 1970 authorized the Environmental Protection Agency (EPA) to establish and enforce national emission standards through the Clean Air Act. National Ambient Air Quality Standards[1] have been promulgated and recently updated (see Box 3.2) for solid and liquid particulate matter (PM) of primarily anthropogenic origin, including pollutants such as ozone. Similar standards have been promulgated by Canada and many European nations.

Much of the documentation of specific airborne hazardous substances and disease has come from investigations of high exposure in industrial environments. This is a reflection of dose response—there is an increased likelihood of disease with more concentrated and longer term exposure to a hazardous substance. Further, morbidity as the result of hazardous airborne substances may not necessarily affect every exposed individual. The very young or old, the infirm, or health-compromised individuals in a population are more likely to be at risk.

Natural Sources and Transport of Airborne Particulate Matter

Although in situ soil particles are natural and vital for plant growth, they are considered to be contaminants when they are entrained in air (or water). Desertification, a significant consideration when contemplating future global climate change, occurs when the reduction of water in the environment, aided by winds, results in mobilization of particles into the atmosphere. The hundreds of millions of tons of soil particles in the lower atmosphere (troposphere) can impact sensitive habitats (e.g., coral reefs

---

[1]See *http://www.epa.gov/ttn/naaqs/standards/pm/s_pm_index.html.*

BOX 3.1
Volcanic Gas Inhalation—Laki, 1783

Iceland sits astride the Mid-Atlantic Ridge, a dominantly submarine mountain chain extending roughly north-south along the midline of the ocean floor. It marks the divergent plate boundary—or rift—between the Eurasian and African plates on the east and the North and South American plates on the west. Basaltic magma wells up semicontinuously along this rift and solidifies as new oceanic lithosphere to form the Mid-Atlantic Ridge; a massive accumulation of lava at the ridge produced Iceland. Active volcanoes are scattered along the plate boundary where it bisects Iceland, and in 1783 an enormous fissure eruption took place at one of these volcanoes. Over an eight-month period, nearly 15 km^3 of lava issued from the Laki rift, covering 580 km^2 in southern Iceland. Concomitantly, volcanic ash, aerosols, and gases were injected into the troposphere and the lowermost stratosphere and were carried by the prevailing westerly winds over Eurasia (see Figure 3.2). The increase in atmospheric albedo resulted in the cooling of northwest Europe by about 1.3°C during the next two years, causing widespread crop failures and famine. A dry fog was observed for more than five months after the eruption. Volcanologists estimate that noxious gases vented to the atmosphere in this event included 122 million tons of sulfur dioxide (SO_2), 7 million tons of hydrochloric acid (HCl), and 15 million tons of hydrofluoric acid (HF).

In Iceland itself, 10,000 deaths were attributed to fluorosis, the result of direct inhalation and biouptake of HF in drinking water, crops, and livestock. From August 1783 to May 1784, the death toll above background levels in England and northern France exceeded 20,000 and 16,000, re-

in the Caribbean; see Kellogg et al., 2004) as well as potentially contribute to an increase in asthma (Prospero, 2001).

An important source of airborne particulate matter is loess, a fine-grained clay-silt material typically derived from glacial comminution of the parent rock. Very large accumulations of loess occur in specific areas of the northern hemisphere, particularly in west-central China and in the U.S. Midwest (Derbyshire et al., 1998), and these areas are a major source of atmospheric particulate matter. The farming practices common during the 1930s in the U.S. Midwest and Southwest, combined with a long period of drought, decreased soil moisture sufficiently to allow winds to generate thick clouds of dust and cause the famous dustbowl. Other natural sources of airborne particulate matter include dune fields and volcanic ash (e.g., ash from the 1980 Mount St. Helens eruption). Maps showing

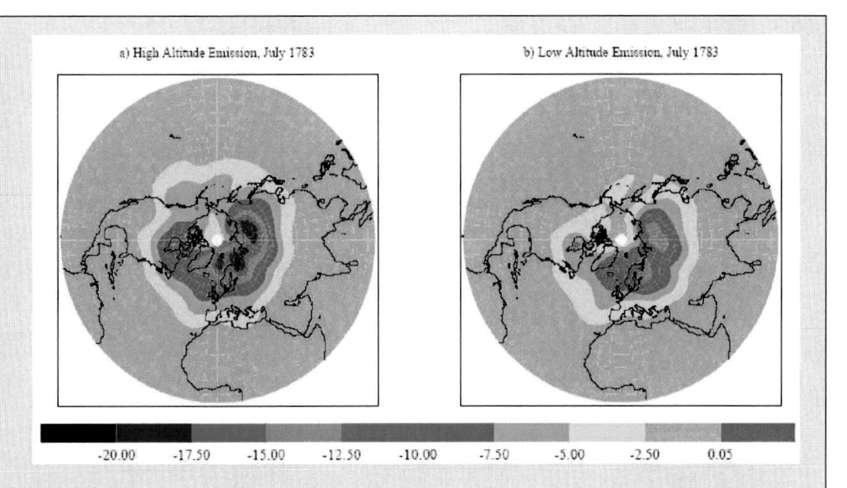

FIGURE 3.2 Model outputs showing the geographic distribution of aerosols from the 1783 Laki eruption. Figures show mean direct radiative forcing (in Wm^{-2}) for high-altitude and low-altitude simulations, relative to a "clean" preindustrial atmosphere. SOURCE: Highwood and Stevenson (2003).

spectively. These figures are incomplete and do not take into account neighboring areas of Europe. Accordingly, the mortality attributable to toxic volcanic gas inhalation (SO_2 and HCl as well as HF) shows that the Laki eruption represents the third most devastating volcanic event in recorded history—after the 1815 and 1883 eruptions of Tambora and Krakatoa (Stone, 2004).

the distribution of bedrock and soil types provide a scientific basis for predicting risk arising from airborne natural particles.

Particle size is important when assessing human health risk because small particles can remain suspended for lengthy periods and thus may be transported great distances, thereby posing a threat to human health through respiratory intake and deposition in nasal and bronchial airways. Sand-sized (0.05–2 mm) and silt-sized particles (2 μm to 0.05 mm), dominated by minerals such as quartz, feldspar, and mica, remain suspended for only short periods of time. In contrast, clay-sized (< 2 μm) particles, which may be any one of the many weathered products of these primary minerals, can remain in air for longer periods of time. In addition, clay-sized and clay mineral particles are normally negatively charged and can sorb positively charged molecules, thus acting as a carrier of associated

BOX 3.2
National Ambient Air Quality Standards for Particulate Matter

The EPA has defined two size categories as relevant for estimation of particulate air pollution: PM_{10}, particulate matter with diameters less than or equal to 10 μm and $PM_{2.5}$, particulate matter with diameters less than or equal to 2.5 μm (see EPA, 2006a). The diameter of the aerosol is defined as the aerodynamic diameter, and the amount of exposure time is also critical.

Particle Size	Exposure Period	Maximum Exposure Amount (μg m^{-3})
PM_{10}	24 hours	150
$PM_{2.5}$	24 hours	35
$PM_{2.5}$	1 year	15

molecules (Walworth, 2005). The size of various solid and biological particles (see Figure 3.3) can be compared with respiratory pathway sizes (Figure 2.4) to illustrate why smaller particles can travel farther into the respiratory system, whereas large particles are likely to be expelled by normal protective mechanisms.

Some of the sources of PM_{10} in the United States are shown in Table 3.1. These data indicate that the major sources of outdoor particulate air

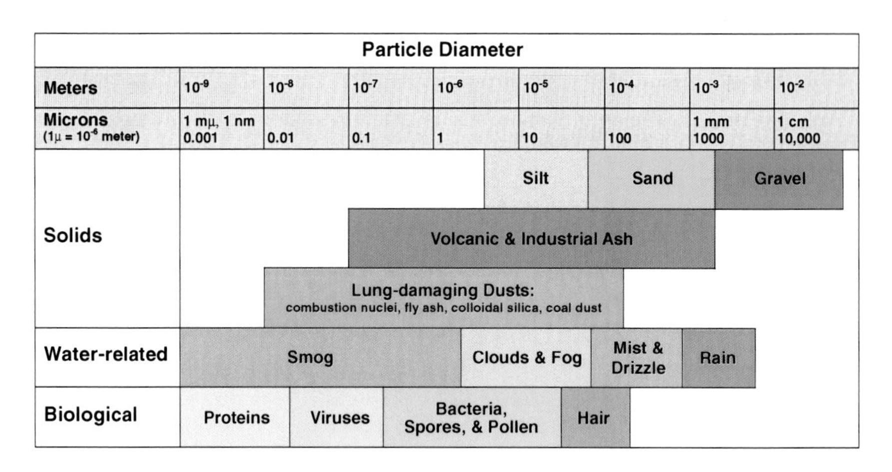

FIGURE 3.3 Comparison of particle size diameters for solids, water particles, and biological airborne materials.
SOURCE: Modified after NIEHS (2006).

TABLE 3.1 Average Annual Emissions of Particulate Matter (PM_{10}) in the United States for 1995–1998 (in millions of tons), from Natural Sources and Human Activities

Source	PM_{10}
Other Sources	
Unpaved roads	11.905
Wind erosion	4.267
Agriculture and forestry	4.937
Construction	3.950
Paved roads	2.489
Fire and other combustion	1.109
Industrial Processes	
Chemical industries	0.065
Metals processing	0.180
Petroleum industries	0.034
Other industries	0.379
Solvent utilization	0.006
Storage and transport	0.097
Waste disposal and recycling	0.302
Fuel Combustion	
Electric utilities	0.288
Industrial	0.263
On-road vehicles	0.276
Nonroad sources	0.458

SOURCE: Modified from CEQ (2006), based on EPA (1998).

pollution are unpaved roads, agriculture and forestry, wind erosion of denuded land, and the use of gravel and sand in construction. The large amounts of airborne particulate matter that are initiated through erosion—the complex interaction of the physical, chemical, and biological weathering of rocks at the surface of the earth—are predominantly sourced from soils.

Health Effects of Mineral Inhalation

There are no known health benefits offered by the inhalation of particulate material. Inspired mineral dust does not appear to provide beneficial elements that can be absorbed by humans, and inhalation of dust or soil has not been reported to offer any protection against diseases. Unlike occurrences of geophagia (soil ingestion) in some cultures (see Chapter 5), the committee knows of no records of cultures "sniffing" earth materials for any purpose.

More than 380 naturally occurring minerals have the potential to be inspired because they may become airborne, although most do not occur in nature in sufficient concentrations to be a health hazard. This section describes three types of earth-sourced particulate material—volcanic, fibrous mineral, and silica dusts—that illustrate the adverse health effects that can be caused by inhaling earth materials.

Volcanic Particles

Dust and aerosol clouds of volcanic origin can dramatically affect human health and welfare. The Mount St. Helens eruption of May 18, 1980, not only illustrated the dramatic power of volcanoes but also the health impacts of volcanogenic particulate matter (see Figure 3.4). Total suspended particulate matter levels downwind of the volcano averaged 33,402 $\mu g\ m^{-3}$ and remained above 1,000 $\mu g\ m^{-3}$ for a week, far exceeding the mean ambient level of 80 $\mu g\ m^{-3}$ (Baxter et al., 1983). This caused diminished light over the Pacific Northwest for several days and increased respiratory morbidity in the emergency workers who were exposed to the resuspended ash (Baxter et al., 1983; Bernstein et al., 1986). Deposited particulate matter was mostly PM_{10}, as was also the case for eruptions between 1997 and 1998 at Soufriere Hills, Montserrat, where respirable dust from resuspended materials at 100–500 $\mu g\ m^{-3}$ levels was hazardous to workers (Searl et al., 2002; Horwell et al., 2003). Another devastating volcanic eruption was the Tambora explosion of 1815, in what is now Indonesia. The volcano expelled ash and dust into the stratosphere, where it remained for several months, resulting in reflection of sunlight back to space which caused a 0.7°C cooling of earth's climate (Matthias et al., 2006). Some 92,000 deaths resulted from the eruption itself and from crop failures and famine in North America and Northern Europe during 1816, the "year without a summer."

Fibrous Mineral Particles—Asbestos

The group of fibrous minerals collectively known as asbestos has become a highly publicized example of particulate matter that is considered to be hazardous. Because of their fibrous morphology and stability at high temperatures, several asbestos minerals have been used as insulation and/or fire retardants, and consequently they have become widely distributed beyond their natural habitats. Excess exposure in occupational environments has led to thousands of studies documenting debilitation, disease, and death.

Although many minerals can occur in a fibrous form, the key factor determining whether fibrous mineral particles are hazardous is their potential to be inhaled. Six naturally occurring minerals—chrysotile, actinolite, amosite, anthophyllite, crocidolite, and tremolite—have been defined

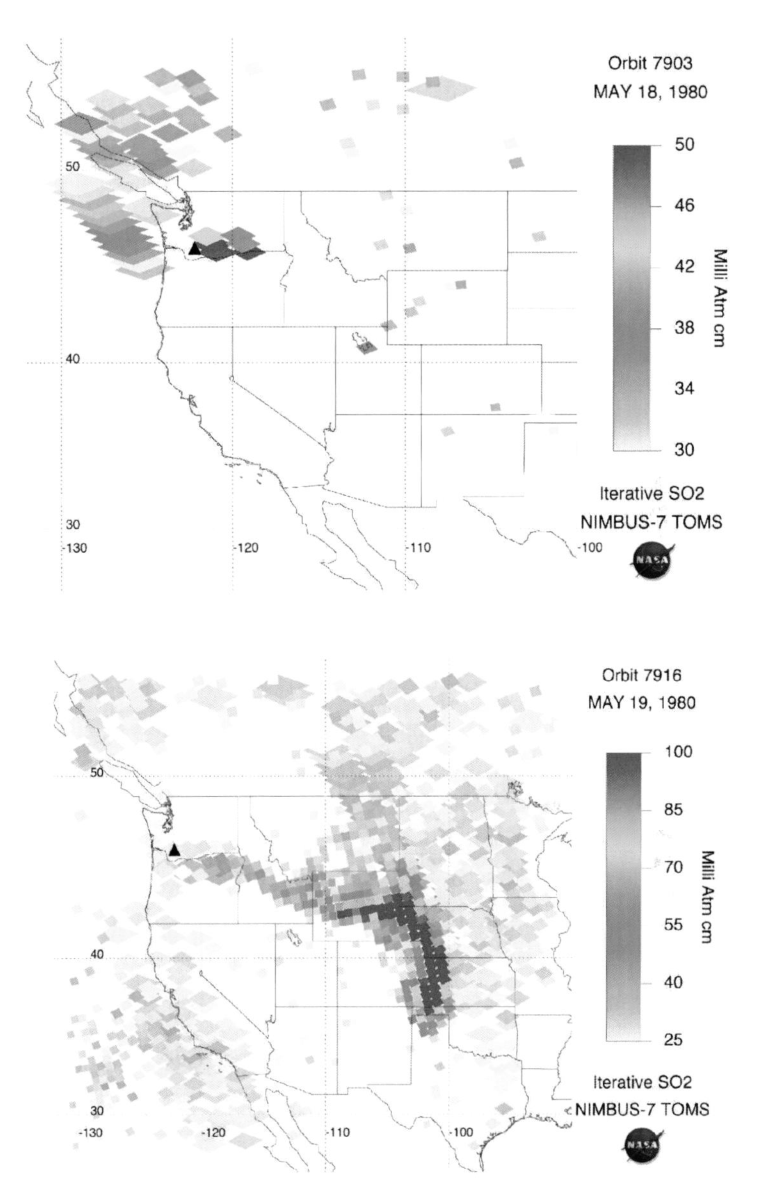

FIGURE 3.4 Satellite imagery showing the spread of sulfur dioxide following the eruption of Mount St. Helens (marked with black triangle) in 1980. The maps display the Sulfur Dioxide Index calculated from data collected by the Earth-Probe TOMS instrument.[2]

[2]See *http://toms.umbc.edu/ and http://toms.gsfc.nasa.gov/*.

TABLE 3.2 Chemical Compositions of Fibrous Minerals Known to Be Hazardous

Mineral Class	Mineral	Chemical Composition
Serpentine	Chrysotile	$Mg_3Si_2O_5(OH)_4$
Amphibole	Actinolite	$Ca_2(Mg,Fe)_5Si_8O_{22}(OH)_2$
	Amosite, var of Grunerite	$(Mg,Fe)_7Si_8O_{22}$
	Anthophyllite	$(Mg,Fe)_7Si_8O_{22}(OH)_2$
	Crocidolite, var of Riebeckite	$NaFe_3^{2+}Fe_2^{3+}Si_8O_{22}(OH)_2$
	Tremolite	$Ca_2Mg_5Si_8O_{22}(OH)_2$
Zeolite	Erionite	$(K_2, Ca, Na_2)2Al_4Si_{14}O_{36}{\cdot}15H_2O$

NOTE: The serpentine and five amphibole minerals are regulated. Erionite, a zeolite mineral, is not regulated but it is recognized as a carcinogen.
SOURCE: NTP (2005).

by the EPA and the Occupational Safety and Health Administration as being hazardous when they have particle length equal to or greater than 5 μm and an aspect ratio (length to width ratio) greater than 3:1. Each of these asbestos minerals (see Table 3.2) has a distinct chemical composition and crystal structure and a different dose-response relationship in humans. The five amphibole minerals are common species found in many rocks and soils, where they can occur as tiny particles or, occasionally, as asbestos fibers. The inhalation potential of these minerals is dependent on particle size (Wylie et al., 1993)—larger particles are less likely to remain airborne after disturbance and consequently are less likely to be respired. Knowledge of the precise crystal structure and elemental composition is crucial for distinguishing between the minerals classified as asbestos and the many other fibrous mineral species (e.g., talc) that may occur with the asbestos minerals. Specific mineral identification is essential for determining potentially hazardous exposure (Skinner et al., 1988, Wilson and Spengler, 1996).

Asbestosis is a noncancerous disease that occurs when the lungs fill with scar tissue (fibrosis—abnormal deposition of the fibrous protein collagen) as a result of continual high exposure to asbestos particulate matter. Fibrosis causes diminished ability to respire essential gases (oxygen, carbon dioxide) and thereby compromises many body reactions. Although ferruginous bodies—nodular aggregates of collagen, mucopolysaccharides, and ferritin—seen during histological examination of lung biopsies provides a telltale signature of particle deposition, high-resolution analytical microscopy is required to accurately identify the particular species of asbestos mineral. Lung scarring may continue after asbestos exposure has ceased. Many workers in occupations where asbestos mineral inhalation may occur also smoke, and respiratory trauma from accumulated insults from different sources increases the risks of contracting disease, especially cancers. Another deadly disease linked to asbestos is mesothelioma, a cancer of the pleura rather than the lung tissues. It may take more than 25 years for this disease to appear in populations (NOHSC, 1999). Mesothelioma is a major public health issue in parts of Turkey, where the responsible fibrous mineral is a zeolite—erionite—rather than an asbestos species (Baris et al., 1979).

Based on detailed assessments of the health effects from occupational asbestos exposure, some members of the amphibole group of minerals are considered to carry a high risk of mesothelioma and lung cancer (Skinner et al., 1988; Ross and Nolan, 2003; IOM, 2006). A recent review (IOM, 2006) noted that asbestos is an established human carcinogen and evaluated a broad range of existing studies to determine whether there was a causal association between asbestos (considered generally rather than by specific fiber type) and specific cancers. This study concluded that there is sufficient evidence to infer a causal relationship between asbestos and laryngeal cancer but that the evidence is only suggestive for pharyngeal, stomach, and colorectal cancers and is inadequate to demonstrate a causal relationship for esophageal cancer. This report noted that uncertainties in its conclusions were a result of limitations in available evidence, and suggested that research was needed to address the relevance of physical and chemical characteristics of asbestos fibers to carcinogenicity.

Chrysotile is the dominant asbestos mineral used worldwide as insulation in public buildings and schools over the past century. In Asbestos, Quebec, where the general population has been exposed to piles of chrysotile waste accumulated for over 100 years, the level of lung cancer appears similar to other areas of Canada, especially when smoking is taken into account (Camus et al., 1998). Although exposure to the asbestos group of minerals as a result of mining or industrial activities receives the most publicity, there are also many cases where there is potential for exposure from natural occurrences (see Box 3.3).

Mineral identification is a key component of any understanding of the relationship between mineral exposure and health consequences, and it is likely that contradictory observations reported by healthcare professionals stem from an inadequate understanding of the physical and chemical characteristics of the mineral materials. Although the public health issues related to asbestos have been aired (e.g., Liddell, 1997; Ross and Nolan, 2003), the specific mechanisms of fibrogenesis and carcinogenesis related specifically to asbestos exposure are not fully elucidated despite contributions from lung physiologists, pathologists, cell biologists, and special studies by numerous medical research teams.

BOX 3.3
Natural Exposure to Asbestos

Serpentinite, an ultramafic metamorphic rock commonly found in California (where it is the state rock), may contain the mineral chrysotile (also known as "white asbestos"). The abundance of serpentinite, and the possibility that the rock may contain asbestos, has prompted concern regarding potential health hazards. The distribution of serpentinite, and thus potential occurrences of chrysotile asbestos, has been documented by the California Geological Survey (Clinkenbeard et al., 2002). Although the report identifies those regions where serpentinites occur, it does not identify whether asbestiform minerals are present nor, if present, the likelihood that asbestos fibers will become airborne and available for inhalation. Although specific long-term occupational exposure to asbestos (e.g., by shipyard workers, plumbers, steamfitters) may cause lung disease, there is no existing evidence that residential proximity to serpentinite rock is hazardous.

A large body of serpentinite near Coalinga, California, is one of the best known chrysotile asbestos deposits in the United States. Originally a mercury mine, chrysotile was mined at the Coalinga and Atlas Mines from the 1950s until the mid-1970s because it was easily extracted in virtually pure mineral form. Chrysotile was also milled on site, creating respirable airborne mineral dust. The Coalinga Mine and Atlas Mine sites, together with mine tailings dumps and milling areas, were remediated during the 1990s as part of the Superfund National Priority List process. The Clear Creek Management Area, adjacent to the mines and included within the Atlas Mine Superfund Site because of natural chrysotile occurrences, is popular with off-highway vehicle recreation and racing enthusiasts. The use of off-road vehicles for recreation, especially where there is little plant cover, can result in significant dust pollution for both riders and downwind populations. Precise mineralogical data for any airborne particulate matter is essential for determining whether such dust poses a health hazard.

Silica Particles

Silica, and the associated disease silicosis, is another example of an intensively studied health effect caused by inhalation of a specific mineral particulate matter. Silicosis, caused by exposure to crystalline silica, is almost exclusively an occupational disease where the key to understanding respiratory problems lies in the size and morphological characteristics of the nonfibrous particles. Construction workers, especially those who sandblast and use jackhammers, are often at great risk for lung disease because silica materials become airborne (Rosner and Markowitz, 1991).

Silicosis can be readily diagnosed radiographically by identification of focal nodular lesions in the upper lung, a site that distinguishes this disease from asbestosis. In both diseases, lung function may not initially be markedly affected, but with continual exposure the progression of lesions reduces the original pliable lung to stiff fibrotic tissues, and occlusion of the air sacs compromises the transmission of essential gases by the respiratory system. Silicosis was identified as a major health concern in the 1930s, but few new cases have appeared in recent years because of increased attention to hazardous workplace environments. The identification of silica in volcanic emissions has caused some recent concern (Horwell et al., 2003).

GAS INHALATION HAZARDS

There is a large, and expanding, inventory of gaseous-phase aerosols that can potentially cause harm to the human respiratory system. These can be natural or anthropogenically generated and are found outside and indoors (McElroy, 2002). The primary natural outdoor pollutants are hydrocarbons from plant respiration, biogenic gases such as methane (CH_4) and hydrogen sulfide (H_2S), and volcanic gases such as sulfur dioxide (SO_2), H_2S, carbon monoxide (CO), and carbon dioxide (CO_2). Radon, a naturally generated gas, is a major indoor air pollutant.

Health Effects of Volcanic Gas Emissions

Volcanic eruptions produce enormous quantities of gas that, in some situations, can have devastating consequences for surrounding plant, animal, and human life. An explosion on August 12, 1986, from Lake Nyos in western Cameroon caused a 100-meter-high jet of water and CO_2 gas, coinciding with a 1-m drop in lake level. An approximately 50-m-thick mist of water and CO_2 rolled down into the surrounding valley, at speeds of over 50 km per hour, killing 1,700 people through suffocation (Freeth and Kay, 1987). The buildup of CO_2 occurred in the lower portions of the lake because the confining pressure of the overlying water mass caused CO_2 derived from the underlying volcanic source to be dissolved and effectively trapped in the bottom waters. Rainwater displaced some of the bottom waters, leading to reduced confining pressure and explosive gas expulsion. Pipes have been inserted into the bottom of the lakes to allow CO_2 to gradually escape and prevent explosive overturning (Evans et al., 1993).

Volcanogenic gas is also being emitted near Mammoth Mountain, California. After an earthquake swarm in 1989 associated with a moving subterranean magma body, U.S. Forest Service personnel noticed an area where the trees appeared to be dying. In 1990, measurements of gas emissions indicated that both CO_2 and helium were venting from the soil in several areas around Horseshoe Lake (see Figure 3.5). The likely source of the CO_2 was contact heating of limestone-rich rocks by a magmatic intrusion. Peak flow occurred in 1991, with soil gas composition as high as 95% CO_2. The current gas flux is approximately stable, with about 100 acres impacted by emission of approximately 110 metric tons of CO_2 per day in the Horseshoe Lake area (McGee and Gerlach, 1998). The dangers posed by these gas emissions were tragically reinforced in April 2006 when three Mammoth Mountain ski patrol members were killed when they fell through snow into a CO_2-charged pit and were asphyxiated.

Health Effects of Biogenic Gas Emissions

The end result of microbial metabolism is often the generation of gaseous byproducts, including CO_2, CH_4, H_2S, nitrogen (N_2), O_2, hydrogen (H_2), and semivolatile compounds like organic acids (Konikow and Glynn, 2005). Most of these gases occur in the soil zone or in saturated sediments, and soil type, water content, mineralogy, and organic carbon content all directly influence the dominant microbial community and therefore the type of gas produced. In some soil environments the concentrations of CO_2 or CH_4 gases reach very high levels—CO_2 concentrations between 1 and 10% are common in productive soils and CH_4 can be over 10% in water-logged soils.

There are examples of very high concentrations of biogenic gases that directly impact human health. In some coal seams, biogenic methane adsorbs on the coal under high confining pressure, and this gas, known as coal-bed methane, is recoverable as an energy resource. The methane is extracted by pumping down the confining aquifer—the decrease in confining pressure results in desorption of the methane, which can then be extracted. Where the coal outcrops at the land surface, however, the methane can vent directly to the surface environment resulting in high methane concentrations in soils. This can result in tree kills and the accumulation of explosive levels of methane in homes. Several areas east of Durango, Colorado, are impacted by methane venting from coal outcrops, with sufficiently high methane concentrations to force several homes to be abandoned.

In anaerobic groundwater that contains dissolved sulfate, the dominant anaerobic microorganism is normally sulfate-reducing bacteria and the metabolic byproduct is gaseous H_2S. Although rarely present in soils at high enough concentration to be toxic (Konikow and Glynn, 2005), H_2S can reach very high concentrations in water, and gas volatilization is enhanced when heated and discharged in residential showers. Legator et al. (2001) noted a higher incidence of central nervous system and respiratory

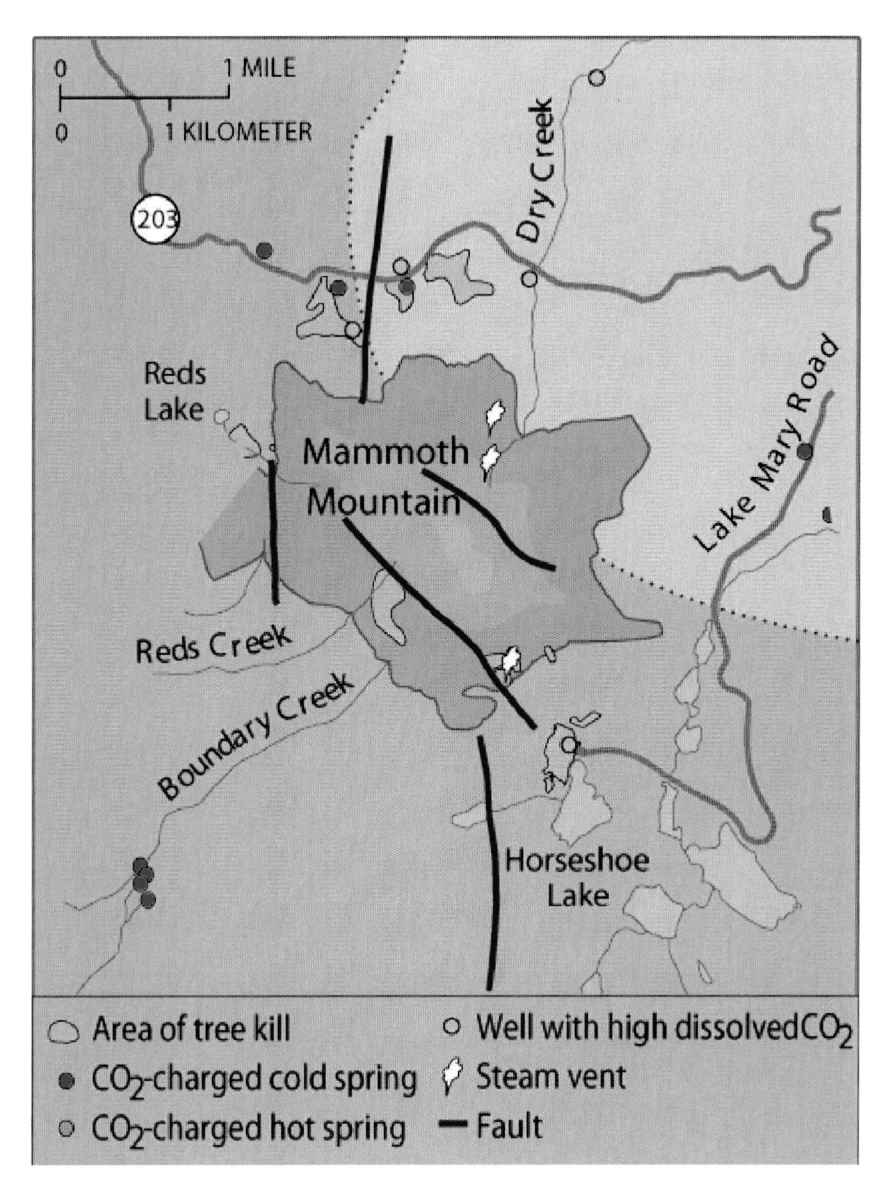

FIGURE 3.5 Distribution of the effects of CO_2 generated by contact heating of limestone-rich rocks by a magmatic intrusion at Mammoth Mountain, California. SOURCE: USGS (2000).

effects as a result of chronic low-level exposure to H_2S. A statistically significant incidence of neurophysiological abnormalities was recorded in a study of the health effects of working and living near a processing plant for high-sulfide oil (Kilburn and Warshaw, 1995). The health effects posed by the outgassing of biogenic or volcanogenic hydrogen sulfide in residential showers is an area of active research and a topic that is an example of research that requires both earth science (to characterize the source of H_2S) and healthcare professionals.

Health Effects of Radon Gas Emissions

Radon is a colorless, tasteless, and odorless gas that is produced by the natural radioactive decay of rocks or soils containing uranium-bearing minerals. Radon is a major contributor to background radiation at the surface of the earth, and ^{222}Rn can be detected in the atmosphere above some geological sites with high radon concentrations. Radon is dangerous because it decays to form radioactive particles that can be inhaled directly or may adhere to dust particles and then be carried into the lungs (Appleton, 2005; Gates and Gundersen, 1992; Graves, 1987). Once inhaled, the short-lived decay products of radon—polonium 218 (^{218}Po), lead 214 (^{214}Pb), bismuth 214 (^{214}Bi), and polonium 214 (^{214}Po)—can become trapped in the lungs and result in lung cancer (Appleton, 2005).

Since 1928, when the International Commission on Radiological Protection advised that low levels of radiation might be harmful, exposure to radioactive elements has been considered a potential carcinogen. Epidemiological studies on humans and experimental studies with animals confirmed the risk associated with high radiation levels from a range of radioactive elements, but not necessarily for low-dose exposures of radon (Friedman, 1988; NRC, 1999a).

High concentrations of uranium (U) and thorium (Th) are present in sedimentary rocks rich in phosphate (phosphorites), in coal beds, and in soils derived from black shales and some relatively reduced granites (Cannon et al., 1978). Geological occurrences of such lithologic units are widespread in the western United States, especially in Montana, Wyoming, and Idaho. Other large deposits occur in Florida, the central Appalachians, and throughout the northeastern part of the country. Locally, gas concentrations are enhanced if the rock mass has fractures or is disturbed by seismic or construction activity. Radon can accumulate in buildings—especially in basements without adequate ventilation—in areas underlain by such rocks. This recently resulted in a radon scare in the northeastern United States which is, in part, underlain by granitic rocks.

The range of radon concentrations in underground mines can be huge, especially when fractured rock permits gas escape. Umhausen, a small town in the Austrian Tyrol that is built on an alluvial fan of a rock slide, had radon measurements between 2,000 and 250,000 Bq m^{-3} (Ennemoser et al., 1994). The incidence of lung cancers was statistically higher than expected in the town population.

INHALATION OF BIOLOGICAL CONTAMINANTS

Biological airborne contaminants—bioaerosols—which can be ingested or inhaled by humans include bacteria, viruses, and fungi of geogenic origin as well as airborne toxins (Griffiths and DeCosemo, 1994). Bioaerosol sizes range typically from 0.5 to 30 μm in diameter, and usually particles are surrounded by a thin layer of water (Stetzenbach, 2001). In other instances, the biological particles can be associated with particulate matter such as soil or biosolids (Lighthart and Stetzenbach, 1994). Bioaerosol particles in the lower spectrum of sizes (0.5–5 μm) are typically of most concern, as these particles are more readily inhaled or swallowed (Stetzenbach, 2001).

Bioaerosols generated from the land application of biosolids may be associated with soil or vegetation, depending on the type of land application, and are therefore considered an earth-derived source of pollution. The soil particles or vegetation provide a "raft" for the biological particles contained within the aerosol (Lighthart and Stetzenbach, 1994). However, for soil particles to be aerosolized, the particles need to be fairly dry, and low soil moisture contents are known to promote microbial inactivation (Straub et al., 1992; Zaleski et al., 2005).

Potentially there are three phases to the bioaerosol exposure pathway—launching of bioaerosols, transport, and deposition onto humans or interception by humans. Launching can result directly from human activity (coughing or sneezing) or indirectly from waste handling and loading of sewage, biosolids, or animal wastes. Launching can also occur from natural sources, such as the wind-blown spores released from soil fungi. Transport distances can be short, as in the case of one human sneezing and infecting a nearby person. In other cases, transport can be over hundreds of kilometers. Human interception of bioaerosols, resulting in infection or illness, can be via ingestion or inhalation.

Health Effects of Airborne Pathogens

Table 3.3 illustrates the wide variety of human pathogens that can be aerosolized. Although most of these pathogens do not originate or reside in soils, there are some significant bacterial pathogens found in soils (e.g., *Bacillus anthracis*, the causative agent of anthrax, although anthrax outbreaks from soil have rarely been documented). Many fungi are found in

TABLE 3.3 Examples of Airborne Pathogens and the Diseases That May Result

Pathogens	Human Diseases
Bacterial Diseases	
Mycobacterium tuberculosis	Pulmonary tuberculosis, disseminated tuberculosis
Chlamydia psittaci	Psittacosis (pneumonia)
Bacillus anthracis	Pulmonary anthrax
Staphylococcus aureus	Staphylococcus respiratory infection, sepsis, cutaneous infection
Streptococcus pyogenes	Streptococcus respiratory infection, sepsis, other streptococcus infections
Legionella spp.	Legionellosis
Neisseria meningitidis	Meningococcal infection, meningitis
Yersinia pestis	Pneumonic plague, bubonic plague
Bordetella pertussis	Pertussis (whooping cough)
Corynebacterium diptheriae	Diphtheria
Fungal Diseases	
Aspergillus fumigatus	Aspergillosis
Blastomyces dermatiridi	Blastomycosis
Coccidioides immitis	Coccidioidomycosis (valley fever)
Cryptococcus neoformans	Cryptococcosis
Histoplasma capuslatum	Histoplasmosis
Nocardia asteriodes	Nocardiosis
Viral Diseases	
Influenza viruses	Influenza
Hantavirus	Hantavirus pulmonary syndrome
Coxsackievirus, Echovirus	Pleurodynia (chest wall pain), respiratory and other infections
Rubivirus	Rubella (German measles)
Morbillivurus	Measles
Rhinoviruses	Common cold
Protozoan Diseases	
Pneumocystis carinii	Pneumocystosis (Pneumocystis Carinii Pneumonia (PCP))

SOURCE: Adapted from Artiola et al. (2005).

soils, including *Coccidiodes immitis*—the causative agent of valley fever—which is prevalent in parts of California and Arizona. In addition to soil-borne pathogens, pathogens can be added to soils via land application of animal wastes or biosolids. Such pathogens include *E. coli* 0157:H7, *Cryptosporidium parvum*, *Salmonella*, enteroviruses, and Norwalk viruses. Although another study (NRC, 2002a) identified a risk of infection to residents living close to land application sites from bioaerosols, analysis of the annual community risk of infection from Coxsackie virus A21 using the one-hit exponential model (Brooks et al., 2005a, 2005b) indicated that community risks from bioaerosols generated during land application of biosolids are very low because of the natural attenuation of pathogens due to environmental factors such as dessication and ultraviolet light.

Health Effects of Airborne Toxins

Endotoxin, also known as lipopolysaccharide, is ubiquitous throughout the environment and may be one of the most important human allergens (Sharif et al., 2004). Endotoxin—derived from the cell wall of gram negative bacteria—is continually released during both active cell growth and cell decay (Brooks, 2004). In soils, bacterial concentrations routinely exceed 10^8 per gram, and soil particles containing sorbed microbes can be aerosolized and act as a source of endotoxin. Farming operations (e.g., driving a tractor across a field) have been shown to cause exposure to high levels of endotoxin. When inhaled, endotoxin can cause a wide variety of health effects, including fever, asthma, and shock (Methel, 2003).

Mycotoxins are secondary metabolites produced by fungal molds. Fungi, such as species of *Aspergillus, Alternaria, Fusarium,* and *Penicillium,* are common soilborne fungi capable of producing mycotoxins. Aflatoxin, produced by *Aspergillus flavus,* is one of the most potent carcinogens known and is linked to a variety of health problems (Williams et al., 2004).

Health Effects of Aeroallergens

Allergic diseases are the sixth leading cause of chronic illness in the United States, affecting roughly 17% of the population. Approximately 40 million Americans suffer from allergic rhinitis (hay fever), largely in response to common aeroallergens, and asthma prevalence and deaths in this country rose substantially from 1984 to the late 1990s (AAAAI, 2000; CDC, 2002, 2004; Mannino et al., 1998; IOM, 2000). Aeroallergens may also contribute to chronic obstructive pulmonary disease and cardiovascular disease (Brunekreef et al., 2000).

OPPORTUNITIES FOR RESEARCH COLLABORATION

The biouptake and coupled interactions of airborne materials—both particulate matter and gaseous components—have important negative impacts on human well-being. In an era when our ability to apply multispectral and hyperspectral satellite data to better understand the nature and characteristics of airborne pollutants continues to increase, our understanding of the sources of airborne pollutants emanating from the surface of the earth, and transported by earth processes, remains inadequate. Even the basic framework for describing the nation's surface—a detailed and comprehensive map of the geochemical and textural characteristics of soils throughout the United States—remains an unrealized high-priority requirement for understanding and predicting risk. And although the detrimental health effects from some natural fibrous and asbestiform minerals have received considerable publicity over the past several decades, there is still inadequate understanding of the precise mineral species characteristics that impact human health.

Collaborative research by earth and public health scientists will be required to effectively address a range of important issues associated with airborne mixtures of pathogens and chemical irritants:

- Exposure concentrations and dose response arising from particulate matter/microbe/chemical interactions.
- Dose response of soil microbes and pollen.
- Long-term risks from low-level concentrations of airborne particulate matter contaminants.

Pollution by wind-blown dusts and volcanic aerosols, gases, and ash is ubiquitous, and most scientists and public health officials predict that the worldwide urbanization phenomenon, combined with the expected effects of global climate change, will generate more potentially hazardous "dusts." A complicating factor is that in most cases natural and anthropogenic air pollution consists of complex mixtures of chemical and biochemical species as well as pathogens, and the earth-sourced or earth-hosted component can be difficult to assess. Adverse effects arising from the inhalation of these species and mixtures require detailed geologic investigations of earth sources and the identification of atmospheric pathways to sites of bioaccessibility and potential ingestion by human hosts. The anticipation or prevention of air pollution–caused health effects prior to the onset of illness requires quantitative knowledge of the geospatial context of disease vectors. A combination of earth observations, using satellite and ground-based detection systems, and public health surveillance has significant potential to improve human health.

4

What We Drink

Water is essential to life and good health. Drinking water provides necessary hydration and serves as the carrier for a variety of substances, both beneficial and harmful, that enter and may be metabolically transformed by the body. The availability and quality of fresh drinking water are controlled by earth and atmospheric processes that generate the water cycle (e.g., NRC, 2004a, 2004b, 2004c). Water that falls on the land surface and does not become part of glaciers or polar ice caps eventually partitions into water that infiltrates into the ground and runoff, which moves by surface or shallow subsurface flow to lakes and streams. The rates of water infiltration are controlled by surface soil textures. The infiltrated water, in turn, partitions into water held in shallow soils, from which it can be removed by plants, and water that percolates downward to the water table where it becomes "recharge" to groundwater. Groundwater moving through geological layers composed of permeable materials (aquifers) eventually migrates to locations at or near the land surface, where it either becomes available for loss to the atmosphere by evapotranspiration or discharges to either fresh surface water bodies or the ocean (e.g., Dingman, 2002). The rates of flow and residence time of groundwater are influenced by the characteristics of the geological layers (Brusseau and Tick, 2006).

At deeper levels of the earth's crust, water occupies the interstices (void spaces) in porous rocks such as fractured lava flows and sandstones, siltstones, and shales (mudrocks). These layered formations may accumulate in depositional basins to thicknesses of tens of kilometers, and consequently the interstitial water contained in such rocks is subjected to in-

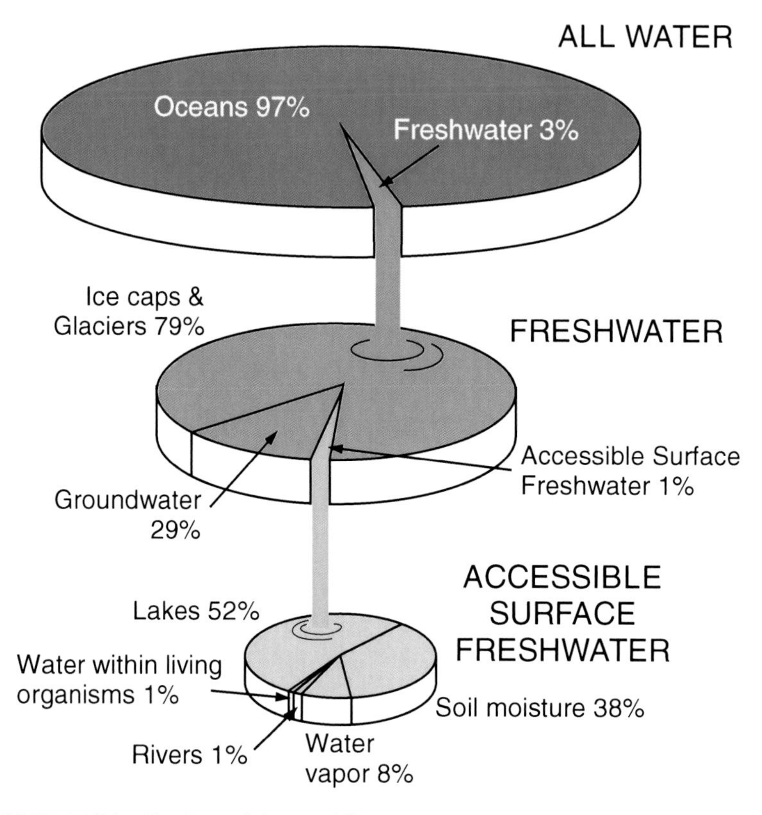

FIGURE 4.1 Distribution of the world's water.
SOURCE: Courtesy "Earth Update" CD-ROM, Rice University and the Houston Museum of Natural Science; used with permission.

creased pressures and temperatures due to burial and compaction. The increased pressure and temperature result in enrichment of aqueous solutions within the rocks by soluble chemical species, especially salts. Such interstitial brines may be several times saltier than the world's oceans, and thus fresh groundwater is typically limited to near-surface reservoirs.

Within the earth's hydrosphere, fresh water comprises only 3% of the total water in the earth system. Because most fresh water is held in glaciers and polar ice caps, only ~30% of fresh water reserves are available as surface water or groundwater for human use (Dingman, 2002; see Figure 4.1). In many arid areas of the world, and even in some more humid locations, groundwater extraction rates by humans exceed natural recharge rates, and the available water stored in aquifers is decreasing. Agriculture is the main user of fresh water, amounting to approximately 70% worldwide and up to 90% in developing countries. Together, irrigation and drainage (especially drainage water reuse) are a major source of salts and toxic trace elements in arid and semiarid regions.

Water, in both quantity and quality, is inextricably linked to public health. The availability of safe water is a basic human necessity and often is the first resource to become critically short in a natural disaster (such as recently encountered along the U.S. Gulf Coast in the aftermath of Hurricane Katrina). The availability and sustainability of safe water from surface and underground sources, particularly in the context of climate changes and population pressure, is a critical area of research that is beyond the scope of this report. This chapter will be limited to a description of the constituents in drinking water as potential benefits to public health (e.g., fluoride) and as potential hazards to public health (e.g., microbial contamination or dissolved toxic elements). The chapter will focus on the threshold research areas, with particular attention to the aspects of water and health that are directly influenced by earth science and the geological framework.

Drinking water contains a variety of substances that result from interactions with geological materials or from other sources such as atmospheric deposition, land application of fertilizer and wastes, mine drainage, and discharge of waste to surface water bodies. These include metals, major and trace elements, natural and anthropogenic organic substances, and microorganisms. Some of these constituents are essential nutrients; many have unknown or only suspected health benefits; and others are clearly health hazards.

In groundwater, inorganic constituents are transported primarily in a dissolved or nanoparticulate form. These constituents are able to enter the drinking water distribution system unless the water is subjected to appropriate treatment processes. Natural organic matter is also a component of natural waters, with largely unknown direct health implications. These substances have a well-established ability to form complexes with metals and potentially enhance the dissolution of minerals and mobilize sparingly soluble metal ions (Hem, 1985). In addition, natural organic matter interacts with chlorine and other drinking water disinfectants to form a dilute mixture of disinfection byproducts that may be mutagenic and/or carcinogenic (Jolley et al., 1984; NRC, 1987; Gerba et al., 2006). Anthropogenic organic materials have a wide variety of potential source points and a wider range of potential health impacts. For virtually all anthropogenic organic compounds, the geological framework—including surface topography, soils, and the vadose zone—exerts a fundamental control on the transport properties from source point to receiver.

HEALTH BENEFITS OF WATERBORNE EARTH MATERIALS

Some beneficial elements, such as calcium, magnesium, and fluoride, either occur naturally in water at sufficiently high concentration to positively influence human health or can be added to water as supplements. In addition, some microbes result in remediation of waterborne contaminants (bioremediation).

Calcium and Magnesium (Hard Water)

Calcium (Ca) and magnesium (Mg) are two of the three most abundant cations (along with sodium) in natural waters. These cations result from dissolution of a variety of rock-forming minerals, including feldspars, carbonates, and sulfate evaporites such as gypsum. Water "hardness" is a measure of the combined calcium and magnesium concentrations in water (see Figure 4.2). Because high concentrations of Ca and Mg generate residues when used with soaps, and boiler scale when water is heated and evaporated, "softening" by ion exchange is recommended for

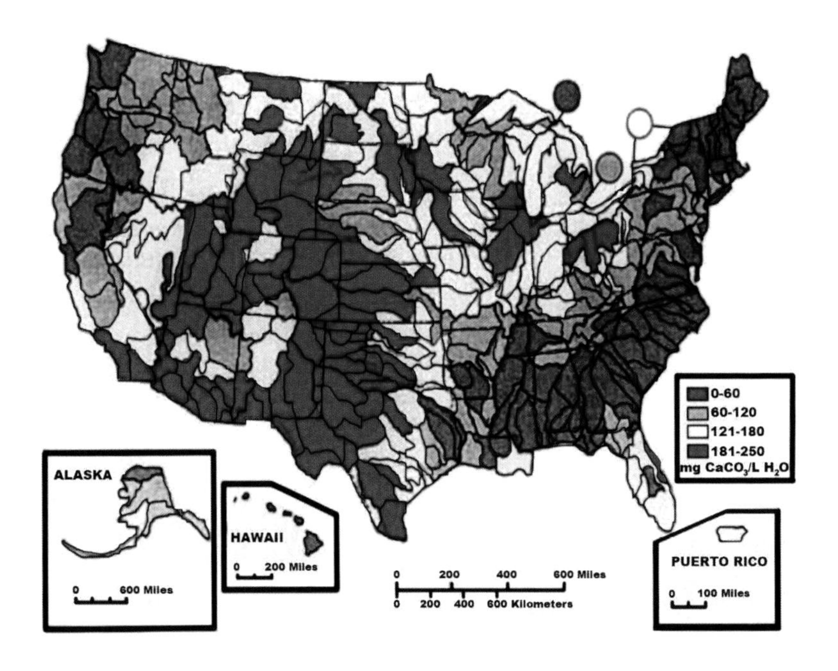

FIGURE 4.2 Water hardness across the United States, 1975 (milligrams per liter). SOURCE: USGS web product; http://water.usgs.gov/owq/map1.jpeg.

npairments from diminished mobility. A recent analysis con-
EPA's drinking water standard of 4 mg L^{-1} is too high to pro-
adverse health effects (NRC, 2006a). Because of this potential
sitive and detrimental health effects, there has been consider-
h by earth scientists to determine the range of fluoride con-
in natural waters and to elucidate the processes controlling
emical cycling of fluorine (Edmunds and Smedley, 2005). Con-
in surface waters are generally much lower than the range of
L^{-1} that promotes dental health (NRC, 2006a), and typically
abundance of fluoride in surface waters does not result in a
enefit. Exceptions are lakes and rivers in volcanic areas where
bodies may receive acidic geothermal fluids containing high
ons of dissolved fluoride. The large ranges of fluoride concen-
groundwater partially result from the variability of fluorine
ons in geological materials and partially because dissolved cal-
free fluoride by precipitating minerals that incorporate fluo-
ry mineral sources in igneous rocks include biotites, amphib-
e, and natural fluorite also occurs in hydrothermal veins. Soils
n fluoride from the parent rock material as well as from an-
c inputs, particularly phosphate fertilizers and sewage sludge.
oncentrations are highly variable on a local scale, due to the
terogeneity of geological materials and the variations in flow
residence times that control water-rock interactions. Accord-
ogical and hydrogeological expertise is essential for both water
grams in areas of potentially high fluoride concentrations and
orts designed to relate health effects, such as skeletal fluorosis,
exposure.

Microbes—Natural Attenuation

gh waterborne pathogens are a globally important health prob-
low), beneficial microorganisms in ground and surface waters
ındamental part of the biogeochemical cycling of elements and
enic organisms, and these can be responsible for the rapid deg-
f a range of organic contaminants. Contamination of shallow
petroleum products, including both gasoline and crude oil, is
d both in this country and around the world. During past de-
ing pipelines, transportation accidents, leaking tanks, and well-
ge resulted in fears of a pending environmental catastrophe.
ata were collected during site investigations and cleanup, how-
s noted that in many cases the extent of the contaminant plume
g from the floating pool of hydrocarbon rarely extended farther
150 m from the source, no matter how big the source was or

water with a hardness exceeding 80 mg L
have led to widespread monitoring of v
countries such as the United States.

There are documented beneficial heal
cium and magnesium, and some calciur
daily water intake. Calcium is particularly
osteoporosis, with a recommended daily
menopausal women of 1,200 mg (Wilkins
cium and magnesium in drinking water
cardiovascular system is controversial—al
ies have suggested an inverse association
cardiovascular mortality, a recently publi
demonstrated no effect of calcium and ma
water on the occurrence of myocardial infa

Magnesium plays an important role as
enzymatic reactions, and the recommended
an adult is about 300–400 mg (SCF, 1993; N
ciency increases the risk to humans of dev
conditions, such as vasoconstriction, hyper
atherosclerotic vascular disease, acute myo
in pregnant women, possibly type II diabete
(Rude, 1998; Innerarity, 2000; Saris et al., 200
ported in multiple clinical and epidemiolog
been supported by the results of experimenta
et al., 2001).

Fluoride

Fluoride, as it occurs in drinking water, h
helps prevent dental caries, particularly in chi
bone mineralization and bone matrix integrity
adding fluoride to water that has low natura
recognized internationally, with community wa
implemented in at least 60 countries. There h
ments of the benefits and risks of fluoridatic
mended the continued use of fluoride in the Uni
tal caries and advocated continued support for
water (HHS, 1991). The report also recommend
tific assessments to determine the optimal expos
all sources (i.e., combined exposure, not only frc
Excess fluoride consumption beyond recomi
may cause fluorosis (a condition that results in st
as well as skeletal deformities that can sometime

functional
cluded th
tect again
for both p
able rese
centratio
hydroge
centratio
0.7—1.2 m
the natu
net healt
these wa
concentr
trations
concentr
cium lin
rine. Pri
oles, ap
may co
thropog
Fluorid
inheren
paths a
ingly, g
supply
researc
to fluo

Al
lem (s
are als
nonpa
radati
aquife
wides
cades
head
As m
ever,
origi
than

water with a hardness exceeding 80 mg L^{-1}. These practical considerations have led to widespread monitoring of water hardness in industrialized countries such as the United States.

There are documented beneficial health effects from the intake of calcium and magnesium, and some calcium and magnesium comes from daily water intake. Calcium is particularly important in the prevention of osteoporosis, with a recommended daily allowance for peri- and postmenopausal women of 1,200 mg (Wilkins and Birge, 2005). Whether calcium and magnesium in drinking water have a beneficial effect on the cardiovascular system is controversial—although many geographic studies have suggested an inverse association between water hardness and cardiovascular mortality, a recently published large case control study demonstrated no effect of calcium and magnesium intake from drinking water on the occurrence of myocardial infarction (Rosenlund et al., 2005).

Magnesium plays an important role as an activator of more than 300 enzymatic reactions, and the recommended daily magnesium intake for an adult is about 300–400 mg (SCF, 1993; NRC, 1997). Magnesium deficiency increases the risk to humans of developing various pathological conditions, such as vasoconstriction, hypertension, cardiac arrhythmia, atherosclerotic vascular disease, acute myocardial infarction, eclampsia in pregnant women, possibly type II diabetes mellitus, and osteoporosis (Rude, 1998; Innerarity, 2000; Saris et al., 2000). These relationships—reported in multiple clinical and epidemiological studies—have recently been supported by the results of experimental studies on animals (Sherer et al., 2001).

Fluoride

Fluoride, as it occurs in drinking water, has two beneficial effects. It helps prevent dental caries, particularly in children, and it contributes to bone mineralization and bone matrix integrity (ADA, 2005). The value of adding fluoride to water that has low natural fluoride levels has been recognized internationally, with community water fluoridation programs implemented in at least 60 countries. There have been periodic assessments of the benefits and risks of fluoridation. A 1991 report recommended the continued use of fluoride in the United States to prevent dental caries and advocated continued support for fluoridation of drinking water (HHS, 1991). The report also recommended the initiation of scientific assessments to determine the optimal exposure level of fluoride from all sources (i.e., combined exposure, not only from drinking water).

Excess fluoride consumption beyond recommended levels, however, may cause fluorosis (a condition that results in striations in tooth enamel) as well as skeletal deformities that can sometimes be severe, resulting in functional impairments from diminished mobility. A recent analysis concluded that EPA's drinking water standard of 4 mg L^{-1} is too high to protect against adverse health effects (NRC, 2006a). Because of this potential for both positive and detrimental health effects, there has been considerable research by earth scientists to determine the range of fluoride concentrations in natural waters and to elucidate the processes controlling hydrogeochemical cycling of fluorine (Edmunds and Smedley, 2005). Concentrations in surface waters are generally much lower than the range of 0.7–1.2 mg L^{-1} that promotes dental health (NRC, 2006a), and typically the natural abundance of fluoride in surface waters does not result in a net health benefit. Exceptions are lakes and rivers in volcanic areas where these water bodies may receive acidic geothermal fluids containing high concentrations of dissolved fluoride. The large ranges of fluoride concentrations in groundwater partially result from the variability of fluorine concentrations in geological materials and partially because dissolved calcium limits free fluoride by precipitating minerals that incorporate fluorine. Primary mineral sources in igneous rocks include biotites, amphiboles, apatite, and natural fluorite also occurs in hydrothermal veins. Soils may contain fluoride from the parent rock material as well as from anthropogenic inputs, particularly phosphate fertilizers and sewage sludge. Fluoride concentrations are highly variable on a local scale, due to the inherent heterogeneity of geological materials and the variations in flow paths and residence times that control water-rock interactions. Accordingly, geological and hydrogeological expertise is essential for both water supply programs in areas of potentially high fluoride concentrations and research efforts designed to relate health effects, such as skeletal fluorosis, to fluoride exposure.

Microbes—Natural Attenuation

Although waterborne pathogens are a globally important health problem (see below), beneficial microorganisms in ground and surface waters are also a fundamental part of the biogeochemical cycling of elements and nonpathogenic organisms, and these can be responsible for the rapid degradation of a range of organic contaminants. Contamination of shallow aquifers by petroleum products, including both gasoline and crude oil, is widespread both in this country and around the world. During past decades, leaking pipelines, transportation accidents, leaking tanks, and wellhead leakage resulted in fears of a pending environmental catastrophe. As more data were collected during site investigations and cleanup, however, it was noted that in many cases the extent of the contaminant plume originating from the floating pool of hydrocarbon rarely extended farther than about 150 m from the source, no matter how big the source was or how fast the groundwater was moving (Mace et al., 1997). While many processes contribute to the fate and transport of hydrocarbons in groundwater, the overwhelmingly most important process is the microbial degradation of petroleum compounds. In virtually all aquifers the rate of microbial degradation of the hydrocarbon contaminant by the native microbial consortium is fast enough that, combined with other attenuation mechanisms (dilution, sorption, and volatilization), the plumes were attenuated within ~150 m. This realization substantially changed the remediation strategy for hydrocarbon contamination of groundwater, and the term "natural attenuation" has now entered the lexicon of the environmental professional (NRC, 2000a). However, natural attenuation is not as effective for chlorinated organic solvents such as TCE and perchlorate, where reduced rates of natural degradation allow plumes to travel several kilometers (Brusseau and Tick, 2006).

HEALTH HAZARDS OF WATERBORNE EARTH MATERIALS

Health hazards from drinking water arise from natural or anthropogenic contamination of source waters used for potable use. In particular, contamination of groundwater is dependent on the earth's materials that host the aquifer. The physical properties of the subsurface result in significant differences in the behavior of groundwater compared to that of surface waters. For example, residence times for groundwater range from a few years to hundreds of years or more. Dilution effects, either in water or the atmosphere, are much less significant for groundwater compared to surface water systems. In addition, the absence of light eliminates the possibility of photochemical reactions, a major route of transformation in lakes or streams. The net result is that once groundwater and the subsurface geological units are contaminated, they are very difficult to decontaminate, and therefore pollution prevention is critical for maintaining sustainable groundwater resources.

Risk assessment is an essential element of effective management of groundwater resources. There are two components to the risk of pollution from groundwater—groundwater vulnerability and contaminant load. Groundwater vulnerability is the intrinsic susceptibility of the specific aquifer in question to contamination (see Table 4.1). An aquifer that is close to the surface, overlain by sandy soil, and located in an area with high precipitation rates would clearly be more vulnerable to contamination than an aquifer in an area of low precipitation that is hundreds of meters below ground surface and overlain by clay soils or other relatively impervious material.

Factors involved in the contaminant load are the type of contaminant, the amount of contaminant released, the timescale of release, and the

TABLE 4.1 Factors Affecting Groundwater Vulnerability to Contamination

Factor	Increases Vulnerability	Decreases Vulnerability
Depth to groundwater	Shallow	Deep
Soil type	Well drained (sandy)	Poorly drained (high clay, organic matter content)
Vadose zone physical properties	Preferential flow channels	Horizontal low-permeability layers
Recharge	High precipitation, high infiltration	Low precipitation, low infiltration
Subsurface attenuation processes	Minimal attenuation	Significant attenuation

SOURCE: Brusseau and Tick (2006).

mode of release. The pollution potential of a contaminant is controlled by its transport and fate behavior. Transport of contaminants from the source zone to groundwater necessitates travel through the soil and vadose zone, where attenuation processes such as sorption and biodegradation can act to reduce and limit such transport. For this reason, the soil and vadose zone are often referred to as a "living filter." The degree to which contaminants will be attenuated is a function of the type of contaminant and the nature of the subsurface (NRC, 2000a). Generally, the greater the amount of contamination released, the greater the pollution potential, although the timescale and mode of release can also affect pollution potential. For example, releases from buried storage tanks may be more prone to cause groundwater contamination than releases from tanks stored above ground on concrete pads.

The greatest groundwater pollution risk is associated with locations where the aquifer has a high vulnerability and the contaminant loading is also high. There are several classes of contaminants within each of the major categories of chemical contaminants: organic, inorganic, and radioactive (see Table 4.2).

A variety of substances associated with human activity can contaminate surface and subsurface water supplies, many of which are virtually unaffected by the geological framework. Nitrate, for example, is an important contaminant in water (see Box 4.1)—principally derived from sewage or fertilizer contamination (NRC, 1995)—that does not significantly

TABLE 4.2 Examples of Organic, Inorganic, and Radioactive Groundwater Contaminants

Organic Contaminants
Petroleum hydrocarbons (fuels)—benzene, toluene, xzylene, polycyclic aromatics, methyl tertiary butyl ether (MTBE)
Chlorinated solvents—trichloroethene, tetrachloroethene, trichloroethane, carbon tetrachloride
Pesticides—DDT (dichloro-diphenyl-trichloro-ethane), 2,4-D (2,4-dichlorophenoxy-acetic acid), atrazine
Polychlorinated biphenyls (PCBs)—insulating fluids, placticizers, pigments
Coal tar/creosote—polycyclic aromatics
Pharmaceuticals/food additives/cosmetics—drugs, surfactants, dyes
Gaseous compounds—chlorofluorocarbons (CFCs), methane, sulfur gases
Agricultural fumigants—methyl bromide
Inorganic Contaminants
Inorganic "salts"—sodium, calcium, nitrate, sulfate, fluoride, perchlorate
Heavy/trace metals—lead, zinc, cadmium, mercury, arsenic
Asbestiform minerals such as crocidolite, chrysotile, and erionite
Radioactive Contaminants
Occurring in solids—uranium, radium, strontium, cobalt, plutonium, cesium
Occurring in gaseous form—radon

SOURCE: Brusseau et al. (2006).

sorb or react with geological materials. In contrast, arsenic and radium (plus radon) are examples of inorganic solutes that have natural earth material sources and significant potential to react with the geological system. These substances have received considerable attention from both the earth science and health science communities, but many uncertainties remain related to sources, exposures, and health effects.

Arsenic

Arsenic is a metalloid element found ubiquitously in nature, occurring in rocks and soil, coal, volcanic emissions, undersea hydrothermal vents ("black smokers"), hot springs, and extraterrestrial material. It is the twentieth most abundant element in the earth's crust, with an average concentration of 2 mg kg^{-1}.

Worldwide, water contamination is the most common source of exposure to environmental arsenic. Currently, the regions of the world with the largest affected populations are Bangladesh and West Bengal in India. High arsenic levels in drinking water have been also reported in Argentina, Chile, China, Colombia, Hungary, Mexico, Peru, Taiwan, Thailand, and parts of the United States (NRC, 1999e, 2001b). In Bangladesh alone (see Box 1.1), it is estimated that more than 95% of the 120 million people

BOX 4.1
The Nitrogen Cycle and Nitrate

The global nitrogen cycle is changing faster than any other major biogeochemical cycle (Townsend et al., 2003). Reactive nitrogen, that is, nitrogen in a chemical form other than N_2, has important worldwide effects on air, water, soil, and human health. Human production of reactive nitrogen is estimated currently to be about 200 teragrams per year (Galloway et al., 2003), and this figure is increasing by about 15 Tg annually. Human addition of reactive nitrogen exceeds worldwide natural production (Fields, 2004). The major contributor is fixation of nitrogen by the Haber-Bosch process (100 Tg), most of which is used as fertilizer and released to the environment. Other human sources are nitrogen-fixing crops such as legumes (40 Tg), burning biomass (40 Tg), draining wetlands (10–20 Tg), and fossil fuel combustion in vehicles and for electrical generation (20 Tg).

Airborne oxides of nitrogen in the troposphere (collectively called NO_x) can produce ozone, which under certain conditions can cause or worsen asthma, cough, reactive airway disease, respiratory tract inflammation, and chronic obstructive pulmonary disease (e.g., Kierstein et al., 2006). At midlatitudes, nitrous oxide (N_2O) behaves as a greenhouse gas, with each molecule being about 200 times as effective in absorbing outgoing infrared radiation as carbon dioxide. In the stratosphere, N_2O catalyzes destruction of ozone. In addition, nitric acid has become an increasingly significant contributor to acid rain in industrial areas of the United States.

Nitrate and nitrite are the two most common forms of reactive nitrogen in water. Both are highly soluble and quickly escape below the root zone into groundwater. Nitrate levels in groundwater in agricultural areas are

living in this region drink tube well water and more than one-third of the tube well water contains arsenic above 50 μg L^{-1} (the guideline value recommended by the World Health Organization is 10 μg L^{-1}; WHO, 2001).

A range of health effects have been associated with long-term chronic arsenic exposure, including cancer (skin, lung, bladder, and kidney), atherosclerosis, and peripheral vascular disease. Epidemiological data have also suggested a link with diabetes mellitus, hypertension, and anemia (Lerman et al., 1980). Many of these studies have been conducted in populations where the exposure to arsenic has been predominantly through contaminated drinking water (WHO, 2004; Smedley and Kinniburgh, 2005).

Although results of national-scale surveys offer a general guide to regions of the United States in which arsenic concentrations in drinking water may be linked to disease (e.g., Figure 4.3), they are not comprehen-

several times greater than in preagricultural groundwater. Increased nitrate loading to coastal areas provides nutrients that cause uncontrollable algae blooms, often rendering the water anoxic and unfit for fish and other oxygen-requiring aquatic organisms (NRC, 2000b).

In humans, ingested nitrate is reduced to nitrite in the saliva. Nitrite is highly reactive, and infants are particularly susceptible. In infants, nitrite combines readily with fetal hemoglobin to inhibit oxygen transport, causing a condition called methemoglobinemia, or "blue baby syndrome," which can lead to death. The international maximum contaminant level of 10 mg L^{-1} for nitrate in drinking water has been set to prevent this condition. The adverse effects of nitrate in infants are exacerbated under some conditions, such as gastrointestinal infection.

In addition, nitrite can combine with commonly occurring secondary amines and amides to form a variety of carcinogenic N-nitroso compounds. Although many vegetables contain nitrate, the presence of vitamin C and other antioxidant compounds can inhibit the formation of N-nitroso compounds.

Epidemiological data relating to cancer risk of nitrate are equivocal. A number of studies have evaluated the geographic distribution of cancer incidence or mortality rates with respect to the distribution of nitrate in water. Associations were observed for gastric and bladder cancers, with uneven findings. Case control and cohort studies, which consider exposure and health outcome on an individual basis, also are inconclusive. To resolve the question of nitrate's carcinogenicity in human populations, it will be necessary to conduct studies in larger populations and include more precise estimates of exposure.

sive and cannot be used to reconstruct doses to individuals. Studies that have examined data from wells over smaller regions in this country have often found large variations in arsenic concentrations that do not correspond to a simple spatial pattern (Ayotte et al., 1999; Peters and Blum, 2003; Root et al., 2005). Local-scale variations in concentration have been detected as a function of the geological formation providing water to wells and the depth of wells within a given formation. In some cases, large temporal variations in a given well have also been observed as a function of pumping conditions. From the earth science perspective, the development of improved understanding not only of the distribution of geological sources of arsenic, but also of the microbial, biogeochemical, and hydrogeological processes that control the mobilization of arsenic from geological materials is a major research challenge.

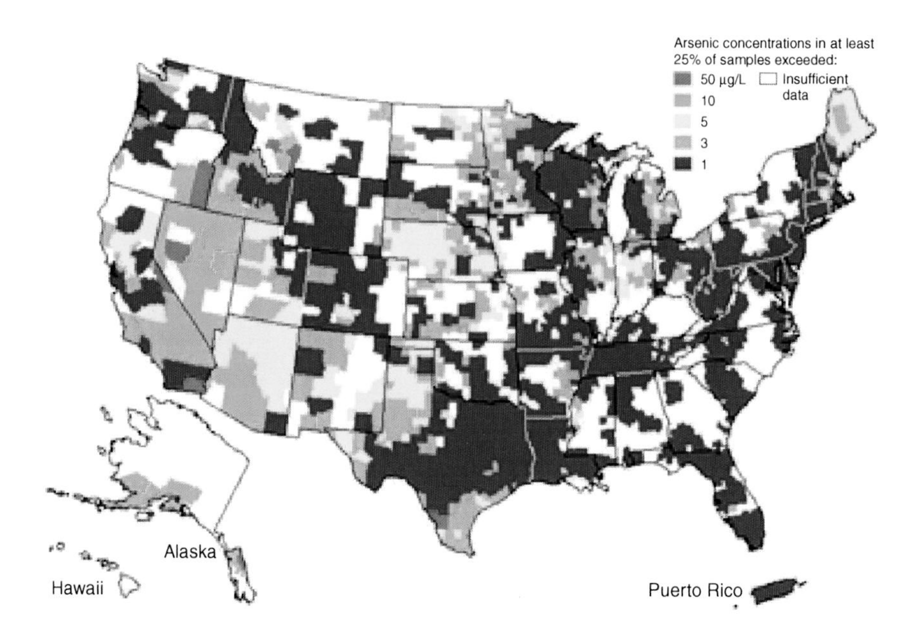

FIGURE 4.3 Arsenic concentrations found in at least 25% of groundwater samples in each county across the United States.
SOURCE: Ryker (2001).

Mercury

Mercury (Hg) in the environment is one of the most widely recognized and publicized pollutants. Natural phenomena, such as erosion of mineral deposits and volcanoes, as well as human activities such as metal smelting, coal-fired electricity generation, chemical synthesis and use, and waste disposal, all contribute to environmental mercury contamination.

Three main forms of mercury occur in the environment—elemental mercury (or quicksilver, Hg^0); inorganic mercury (Hg^{1+}, Hg^{2+}); and organic methyl-, ethyl-, and phenylmercury. Each form has a different solubility, reactivity, and toxicity. Because of biomethylation and bioaccumulation, the effects of methylmercury (MeHg)—the most toxic of the organic forms—vastly exceed those of inorganic mercury as a result of the transport of MeHg across the blood-brain barrier as a complex with L-cysteine (Clarkson, 2002). Mercury is a potent neurotoxin which can cause developmental effects in the fetus as well as toxic effects on the liver and kidneys of adults and children. Over 60,000 babies born in the United States each year are at risk of neurodevelopmental effects from in utero exposure to methylmercury (NRC, 2000c). Sublethal effects of mercury toxicity include diminished ability to learn, speak, feel, see, taste, and move. Children under the age of 15 are most vulnerable because their central nervous systems are still developing.

Although mercury toxicity can occur through skin contact, inhalation, or ingestion, the major route of human exposure to mercury released in the environment is through consumption of contaminated fish. Elevated concentrations of MeHg in fish occur even in the absence of direct anthropogenic discharges of mercury to the water bodies in which the fish live. Natural or geogenic sources of mercury also occur in lakes (Rasmussen, 1996). Elevated MeHg concentrations of fish in remote lakes are generally considered to be influenced by inputs of atmospheric inorganic mercury directly to the lakes and indirectly via their watersheds. Anthropogenic mercury emissions are probable contributors to mercury loading of lakes—these anthropogenic inputs originate as emissions from coal combustion, as waste incineration, and as emissions from other industrial and mining processes. Emissions from coal-fired power plants are a major source of mercury in the atmosphere and hence, by deposition, on land and in water. Volcanic emissions are a natural contributor of mercury to the environment, and recently regional and intercontinental dust movement has been suggested as a source of natural mercury contamination (Holmes and Miller, 2004).

Because of human activities, there has been significant redistribution of heavy metals, including mercury, from areas where they have little impact on human and animal health to areas where they can be detrimental to human health. Significant gaps remain in our quantitative understanding of the mercury cycle; although much of the process that forms MeHg is highly site specific, it is superimposed on a global cycle in which Hg(0) is the principal mercury species. Even basic questions, such as the effect on mercury concentrations in fish if atmospheric mercury deposition is reduced, or the extent of mercury toxicity where there is interaction with trace elements (e.g., selenium), have yet to be addressed. The complexity of the problem will require input from many disciplines, including public health and earth sciences.

Selenium and Molybdenum

Selenium and molybdenum frequently occur together in soils, and these trace elements can be concentrated by agricultural practices, for example, in the San Joaquin Valley in California (Ong et al., 1997). Agricultural irrigation of soils high in selenium and molybdenum results in solubilization, and ultimately bioaccumulation, of these trace elements. Kesterson Reservoir in California is a highly cited example where irrigation water that contained high selenium levels resulted in birth defects and embryonic abnormalities in birds (Ohlendorf et al., 1990). The resulting restrictions on utilization of subsurface drainage have led to innovative technologies to reduce drainage volumes. These include reutilization of drainage water to irrigate salt-tolerant crops or halophytes, so that the volume of drainage water is reduced prior to salt concentration via solar evaporators (Oster and Grattan, 2002). The role of selenium ingested in food is described in more detail in the next chapter.

Radium and Radon

Radioactive contaminants occur naturally in groundwater originating from geological sources or are present in surface water as the result of contamination from a range of sources that include weapons testing, nuclear power plants, landfills, and medical applications. Radioactive concentrations may be several times higher in groundwater than in surface water (NRC, 1999c). Exposure routes for radium and radon include direct consumption of contaminated water and by inhalation (Appleton, 2005); development of lung cancer has been linked to household inhalation of radon (see Chapter 3).

Microbes

Microbial pathogens include bacteria, viruses, and protozoan parasites. Bacteria are common infectious agents implicated in many waterborne disease outbreaks—*Vibrio cholerae, Helicobacter, Campylobacter, Salmonella, E. coli,* and *Shigella* are bacteria known to cause infections and even death. Many bacterial diseases are due to failures of water treatment systems, resulting in consumption of untreated groundwater or surface water (e.g., Gerba and Pepper, 2006). One recent example involved inadequate treatment of the water supply in Walkerton, Ontario, which enabled high levels of a pathogenic strain of *E. coli* 0157:H7 to contaminate the entire drinking water system, resulting in seven deaths and hundreds of illnesses (McIlroy, 2001). Worldwide, waterborne diseases result in millions of deaths each year. Some 2.4 billion people in developing countries still have no access to basic sanitation, resulting in illnesses such as typhoid fever, dysentery, and cholera (WHO and UNICEF, 2000).

The role that viruses play in the waterborne transmission of human diseases is less well understood than that of many bacteria and protozoa, mainly due to the difficulties associated with detecting viruses in water. There are many routes of exposure, including consumption of drinking water polluted by viruses, consumption of shellfish harvested from contaminated water, consumption of food crops grown in soil irrigated with wastewater or fertilized with sludge, and contact with contaminated recreational water. Human enteric viruses are small (25–100 nm) and are encased in a resistant structure that protects them from environmental degradation and disinfection (Sobsey, 1989). Norwalk virus, the most widespread human calicivirus, causes outbreaks of waterborne and foodborne viral gastroenteritis. According to the Centers for Disease Control and Prevention (CDC), more than 96% of reported outbreaks of nonbacterial gastroenteritis characterized by nausea, vomiting, diarrhea, and an illness lasting one to three days are caused by Norwalk virus (Fankhauser et al., 1998), with an estimated 23 million cases per year in the United States (Mead et al., 1999).

Protozoan parasites are eukaryotic organisms that have been implicated as agents of waterborne disease. These include *Giardia lamblia* and *Cryptosporidium parvum* in particular, and the emerging pathogens *Microsporidia* and *Cyclospora*. Another emerging pathogen, *Naegleria fowleri*, is found in both soil and surface water environments. This protozoan parasite infects humans by entering via the nose and subsequently travels to the brain, where it multiplies in the central nervous system and ultimately results in death via primary amebic meningoencephalitis (Zhou et al., 2003). This parasite has recently been detected in 8% of drinking wells in southern Arizona (C.P. Gerba, University of Arizona, personal communication, 2006).

The transport of microorganisms through soils or the vadose zone is affected by a complex array of abiotic and biotic factors, including adhesion processes, filtration effects, soil characteristics, water flow rates, predation, the physiological state and intrinsic mobility of the cells, and the presence of biosolids. Viruses have a large potential for transport, although transport is limited when they adsorb to soil colloidal particles and biosolids. Virus sorption is controlled by the soil pH. The larger size of bacteria means that soil acts as a filter, limiting bacterial transport. Soil should also limit the transport of the even larger protozoa and helminthes. Soils and aquifers in fractured rock and porous media contain zones of preferential flow. Both water and small particles, including inorganic colloids and microorganisms, may be transported through these preferential flow zones and macrochannels very rapidly, and since particles are not subject to diffusion into the finer porosity matrix material, the result is extremely rapid transport. The transport properties of pathogens passing through heterogeneous porous and fractured media and colloid-facilitated transport of contaminants are a critical area of research that will demand expertise in hydrogeology, rock mechanics, geophysics, microbiology, and public health.

Pharmaceutical Substances

There are thousands of organic compounds that can potentially contaminate potable water sources, and many studies have examined the fate and transport of various classes of organic contaminants found in drinking water. Here the focus is on pharmaceuticals, including endocrine disrupting compounds (EDCs)—chemicals that modify the function of endocrine glands and their target organs (NRC, 1999b; Arnold et al., 2006; see Box 4.2)—as an example of an emerging organic contaminant category that occurs in groundwater and is a concern in many countries (Plant and Davis, 2003).

In 1999–2000 the U.S. Geological Survey (USGS) carried out a comprehensive reconnaissance in streams throughout the country that are potentially affected by human activities (Kolpin et al., 2002). Up to 95 trace chemicals frequently present in municipal wastewater were measured at 139 sites. Eighty percent of the waters tested by the USGS contained at least one of the 95 trace contaminants, and 82 of the 95 were present at one or more of the 139 sites (see Table 4.2).

Many studies of the potential for adverse effects from exposure to EDCs in humans and wildlife have been carried out in the United States and in Europe (e.g., NRC, 1999b; Arnold et al., 2006). EDCs in wastewater effluent have adversely affected fish and other wildlife, but there is con-

BOX 4.2
Endocrine Disrupting Compounds

Endocrine disrupting compounds (EDCs) have a number of characteristics:

- EDCs interfere with the synthesis, secretion, transport, binding, action, or elimination of natural hormones in the body that are responsible for the maintenance of homeostasis (normal cell metabolism), reproduction, development, and/or behavior.
- EDCs can be hormone mimics, with hormone-like structures and activities. That is, EDCs sometimes have chemical properties similar to hormones and bind to hormone-specific receptors in or on the cells of target organs.
- EDCs frequently have lower potency than the hormones they mimic (i.e., require a higher dose to elicit an equivalent response) but may be present in water at high concentrations relative to natural hormones. Furthermore, EDCs may not be subject to normal (internal) regulation mechanisms.

TABLE 4.2 Hormones and Hormone Mimics Observed in U.S. Surface Waters

Compound	Description	Detection Limit (μg L^{-1})	Frequency of Detection (%)	Max. (μg L^{-1})	Median (μg L^{-1})
Progesterone	Reproductive hormone	0.005	4.1	0.199	0.11
Testosterone	Reproductive hormone	0.005	4.1	0.214	0.017
17β-estradiol	Reproductive hormone	0.05	9.5	0.093	0.009
17α-estradiol	Reproductive hormone	0.005	5.4	0.074	0.030
Estriol	Reproductive hormone	0.005	20.3	0.043	0.019
Estrone	Reproductive hormone	0.005	6.8	0.027	0.112
Mestranol	Ovulation inhibitor	0.005	4.3	0.407	0.017
19-norethisterone	Ovulation inhibitor	0.005	12.2	0.872	0.048
17α-ethinyl estradiol	Ovulation inhibitor	0.005	5.7	0.273	0.094
cis-androsterone	Urinary steroid	0.005	13.5	0.214	0.017
4 nonylphenol	Detergent metabolite	1.0	51.6	40	0.7
4-nonylphenol monoethoxylate	Detergent metabolite	1.0	45.1	20	1
4-nonylphenol diethoxylate	Detergent metabolite	1.1	34.1	9	1
4-octyphenol monoethyoxylate	Detergent metabolite	0.1	41.8	2	0.15
4-octyphenoldiethoxylate	Detergent	0.2	23.1	1	0.095
Bisphenol A	Plasticizer	0.09	39.6	12	0.13

NOTE: Median concentrations were determined on the basis of samples in which the respective chemicals could be measured; that is, negative results were ignored in arriving at a median concentration.
SOURCE: Pepper et al. (2006).

siderable controversy as to whether human health has also been adversely affected by exposure to endocrine-active chemicals because of inconsistent and inconclusive results. Dose-response relationships are likely to vary for different chemicals and endocrine disrupting mechanisms, and such relationships may be species dependent. The exposure sets that do exist are primarily from chemical levels in various environmental media such as air, food, or water and may not reflect internal concentrations in blood or endocrine-regulated tissues. Exceptions to this are human breast milk and adipose tissue (e.g., Swan et al., 2005). Overall, more research is needed. It remains difficult to assess the risk to human and animal health from endocrine disruptors due to the necessity to extrapolate from low-dose exposures. An additional difficulty in assessing risk from endocrine disruptors is the possible synergistic effect from other environmental hazards. Because controversy will continue to surround EDCs and their potential short- and long-term risks to environmental and human health and welfare, EDCs remain an "emerging issue."

OPPORTUNITIES FOR RESEARCH COLLABORATION

There is a rich array of opportunities for earth and public health scientists to collaborate on research that addresses health and drinking water quality. The earth science component of this research relates to improving the understanding of sources, transport, and transformations of potentially hazardous substances in water to ultimately determine the concentrations to which people are exposed through their drinking water. The health components of the research relate to quantifying and understanding the mechanisms of human responses to these exposures. The overall goal in all cases is to be able to predict potential health effects based on improved process-based understanding and, where appropriate, through modeling. Prediction of potential adverse health effects will provide the basis for development of effective prevention or mitigation measures related to either the water source or the human health response. High-priority collaborative research activities are to:

1. Determine the health effects associated with water quality changes induced by technologies and other strategies currently being implemented, or planned, for extending groundwater and surface water supplies to meet increasing demands for water by a growing world population. Of particular interest with respect to groundwater are changes in water quality induced by:

- changes in rates and locations of groundwater extraction;
- treatment of sewage effluent for potable reuse;

- artificial recharge using stormwater and treated wastewater;
- water "banking" via injection or aquifer storage and recovery;
- extraction of brackish groundwater for desalination;
- introduction of imported or recycled irrigation water; and
- changes in land use, vegetation, and irrigation practices that alter rates of infiltration, drainage water runoff, and evapotranspiration.

All of these have the potential to alter the major and minor ion composition of groundwater. For example, fresh surface water stored in a brackish aquifer during aquifer storage and recovery will experience an increase in total dissolved solids due to mixing with ambient brackish water and dissolution of minerals from the aquifer matrix. Many of these may also introduce contaminants such as microbial pathogens, organic contaminants such as pesticides or solvents, and inorganic contaminants such as nitrates or metals. Of particular interest with respect to surface water are changes in water quality induced by urban and agricultural runoff, discharge of waste effluents from municipal or industrial sources (including the extractive mineral and energy industries), construction and operation of dams and reservoirs, drainage of wetlands, and channel modifications for purposes of flood control, navigation, or environmental improvement.

2. Identify and quantify the health risks posed by "emerging" contaminants, including newly discovered pathogens and pharmaceutical chemicals. The health effects of many naturally occurring substances at low concentrations and the health effects associated with interactions of multiple naturally occurring substances are poorly understood. Public health professionals and earth scientists will need to collaborate to identify emerging substances of potential concern and to improve understanding of the processes controlling the mobility of these substances in the environment, particularly in light of potential changes in concentrations induced by human activities that alter the land or the hydrological cycle. Of particular interest are "emerging" contaminants such as hormones, pharmaceuticals, personal care products, and newly identified microbial pathogens for which sources and transport processes are poorly understood. The synergistic and antagonistic interactions of mixtures of contaminants with naturally occurring substances in water also pose priority research questions. Examples of specific research priorities include an understanding of the:

- fate and transport of prions from soil to groundwater and surface water and their relationship to disease incidence;

- fate and transport of viruses through soil and vadose zones to groundwater and their relationship to disease incidence;
- fate and transport of *Naegleria fowleri* from soil to water and disinfection strategies for contaminated wells;
- fate and transport of endocrine disruptors through soil to groundwaters and the influence of long-term, low-level exposure on human health; and
- fate, transport, and human health effects of perchlorate from soil and groundwater.

5

What We Eat

Public health effects from what we eat are a consequence of both direct (inadvertently or consciously eating earth materials) and indirect (via food) ingestion paths. The latter represents a considerably more important risk to public health, and consequently this chapter has a primary (but not exclusive) focus on soils—the major exposure pathway between earth science and the human health issues that are associated with eating. This chapter reviews the exposure pathways represented by direct ingestion of earth material and by indirect ingestion arising from both microbial activity in soil and the trace elements and metals present in soil and other earth materials.

Knowledge that a link exists between geology, microbiology, and food is as old as our knowledge of soil and agriculture. Minerals, organic material, microorganisms, and dissolved metal species in soils are in close proximity to the roots of food crops. Many factors—the dynamic temporal and spatial variability in chemical speciation and mobility, microbial community structure, pathogen viability, and organic contaminant mobility and persistence—must all be considered when assessing the public health impacts of earth materials. The interrelated topics of foodborne pathogens, the microbiology of food and food spoilage, and agricultural microbiology are enormous and beyond the scope of this report.

EATING EARTH MATERIALS (GEOPHAGIA/GEOPHAGY)

Although soil or clay contamination of foods is not recognized as either an immediate threat or a benefit to human health, the direct consumption of soil or clay *as* food—known variously as geophagia, geophagy, or pica—is a classic example of the intersection of earth science and public health. Human consumption of earth materials has been documented from historical times, and both involuntary and voluntary consumption of soil or clay occurs today (Abrahams, 2003, 2005). Because of immigration, the tradition of geophagia has been introduced and is increasing in Western societies, and imported soils can often be found in local ethnic food stores in this country for sale to immigrants. Geophagia is a potential route for transmission of pathogens (e.g., helminthes, see below) directly to the human host through ingestion of soil (Magnaval et al., 2001; Santamaria and Toranzos, 2003).

Geophagia is considered by many human and animal nutritionists to be either:

- an acquired habitual response in which clays and soil minerals are specifically ingested to reduce the toxicity of various dietary components common to the local environment (e.g., in tropical rain forests, where many plants and fruits contain toxins to reduce their palatability) or
- an innate response to nutritional deficiencies resulting from a poor diet, typically rich in fiber but deficient in magnesium, iron, and zinc. Such diets are common in tropical countries, particularly where the typical diet is dominated by starchy fiber-rich foods such as sweet potatoes and cassava.

From an historical perspective, geophagia has also been commonly associated with various mental disorders and afflictions that have a wide variety of rather unpleasant cures. Even today, the theory of geophagia as a subconscious response to dietary toxins or stress must be balanced against the habitual eating of soil that has been reported to develop into extreme, often obsessive, cravings. These cravings generally occur immediately after rain. Typical quantities of soil eaten by geophagics in Kenya have been reported to be 20 g per day—almost 400 times more than typical quantities of soil thought to be inadvertently ingested through hand-to-mouth contact (i.e., about 50 mg per day) or with leafy vegetables. Although eating such large quantities of soil increases exposure to essential trace nutrients, it also significantly increases exposure to biological pathogens and to potentially toxic trace elements, especially in areas associated with mineral extraction or in polluted urban environments.

HEALTH EFFECTS OF MICROBES IN EARTH MATERIALS

Clearly, the greatest direct benefit of earth materials to public health with respect to what people eat is that surface soils provide a medium for food production, either directly consumed by humans or indirectly consumed via food animals. In either case, plant nutrition is the result of soil characteristics that ultimately affect human health and welfare. Many soil microorganisms aid plant growth and food production. These include free-living microbes in the rhizosphere and symbiotic associations involving rhizobia and mycorhizal fungi. Both of these symbiotic associations improve the nutritional content of plants, contributing nitrogen, phosphorus, and trace elements. In addition, some natural soil microbes suppress plant pathogens (Press et al., 2001; Zehnder et al., 2001), and other soil microbes can remove or transform organic toxicants in soil (see Box 5.1).

The simplest examples of geologically influenced direct microbial threats to human health in food are soilborne human pathogens. Many of the major enteric pathogens are transmitted via the fecal-oral route. Pathogenic organisms in soils can infect and damage either food crops or the animals and humans that ingest them (Tate, 2000). Antibiotic uptake by plants has also been demonstrated, with clear potential to adversely impact human health (Kumar et al., 2005). Some human pathogens are naturally present in soils but do not commonly infect plants as an intermediate host (food spoilage and fermentation are excluded as examples here). The most common mechanism for transmission to humans is from soil adhering to unwashed agricultural products or by transfer of waterborne pathogens introduced during irrigation or food processing (Heinke, 1996; Maier et al., 2000; Tate, 2000). Agricultural uses of treated sewage sludge (biosolids), sewage effluent, or human waste (night soil) are all potential sources of human pathogens in agricultural products. Human pathogens found in these materials include viruses (e.g., coxsackie or poliovirus), bacteria (including *Salmonella* and *E. coli*), and protozoan parasites (e.g., *Giardia* or *Cryptosporidium*) (NRC, 2002a).

Another important group of introduced pathogens commonly present in soils are helminthes (worms—roundworms, flatworms, and tapeworms). Roundworms (nematodes) are the most common helminthes in soils, and these are frequently ingested by people in developing countries and in the southeastern United States. Surveys have demonstrated that 75% or more of the populations in rural areas in Latin America and Africa are infested with intestinal roundworms such as *Ascaris lumbricoides*. In many cases, their presence is asymptomatic, but in other cases, heavy worm burdens can cause anemia, vitamin deficiencies, and blockages of the intestine and common bile duct.

The other direct threat to public health is from plant pathogens that

BOX 5.1
Anthropogenic Contaminants and Natural Attenuation

Soil microbial populations form one component of the "natural attenuation" approach to the remediation of contaminated soil (NRC, 1993, 2000a), where the subsurface microbial community degrades contaminants. This interaction often (but not inevitably) results in a decrease in the concentration of toxic components and eventual remediation of the contaminated soil, providing protection for the food supply.

Growing evidence exists for a link between the geochemical and mineralogical properties of a subsurface system and the efficiency of biotransformation of organic contaminants (Rogers et al., 1998a, 1998b; Rogers, 2000; Rogers and Bennett, 2004; Bennett et al., 2000, 2001). In a typical hydrocarbon-contaminated soil or aquifer, the biogeochemical system is carbon substrate rich but nutrient poor, and the overall attenuation efficiency is limited by available ferric iron, nitrogen, and/or phosphorus (Chapelle, 2001). Laboratory and field experiments have demonstrated that the inorganic nutrient content of the constituent minerals directly influences the rate of hydrocarbon degradation by anaerobic microorganisms. In methanogenic regions, phosphate is the critical nutrient, whereas in iron-reducing zones the availability of nitrogen and iron constitute the limiting nutrients. Soil mineralogy is therefore a key control on the microbial detoxification of soil and a fundamental part of the microbial habitat description.

Another important example of indirect benefits of soil microbes, and the influence of the earth sciences on what we eat, is in the efficient breakdown of the various organic pesticides used to enhance agricultural yields (Corona-Cruz et al., 1999; Ragnarsdottir, 2000). After application, a valuable attribute of an effective pesticide is efficient action on the target plant or insect pest, followed by rapid degradation to limit runoff or indirect damage to valuable organisms (Maier et al., 2000). Both abiotic and biotic mechanisms are critical in degrading organic pesticides, and geological factors are important for both. In particular, soil pH, clay content, and moisture content are important factors for microbial degradation of organic pesticides.

infect agricultural products and decrease yields or damage the products (Tate, 2000). Although the geological controls on direct infection by plant or human pathogenic organisms have not been extensively investigated, the geochemical environment is a fundamental attribute of the soil habitat, and the basic physical attributes of soil such as pH, temperature, moisture content, porosity, permeability, and organic matter content are all important for the viability of pathogenic organisms and the structure and species composition of the soil microbial community (Kodama, 1999). The bulk physicochemical attributes of soil are a direct function of the local geology (e.g., the amount of clay and type of clay mineral present, the carbonate content of the soil, or the availability of mineral iron oxides).

Viral particles behave in a fashion resembling passive charged colloids, where the charges of the viral and mineral surfaces are the important characteristic (Ashbolt, 2004). Viral colloids interact with other charged surfaces, particularly clays, and this interaction can be modeled using well-established double-layer models for solute sorption and exchange on mineral surfaces. Soil mineralogy and mineral surface geochemical properties provide a clear geological component to viral transport.

A recently advanced theory on microbial transport in porous media involves the nutrient requirements of the microbial population and the nutrient content of the soil or aquifer minerals (Rogers and Bennett, 2004). In a sand and gravel aquifer contaminated with crude oil, field and laboratory experiments showed that different minerals support dramatically different microbial populations independent of the mineral surface charge. The primary control of surface attachment and adhesion is the nutrient content of the mineral, with the critical nutrient varying with the dominant metabolic guild. Although physical filtration is the principal factor controlling microbial transport in fine-grained soils, in coarse-grained soils both the mineral surface charge and the mineral chemical composition also influence transport.

Microbial metabolism also represents an indirect threat to public health, through biogeochemical cycling of elements, alteration of soil gas composition, weathering of minerals, and altering element speciation (Chapelle, 2001; Ehrlich, 1996). For many slow geochemical processes, microbial catalysis is the primary mechanism for rapid and significant change in metal speciation and mobility (Huang et al., 2004; Islam et al., 2004). Microorganisms are now recognized as an important factor in agriculturally important metal chemistry, particularly for iron (Burd et al., 2000), as well as in the chemistry of toxic metal contaminants. Basic geological attributes such as mineralogy and permeability directly influence soil pH, moisture content, and redox potential and, as a result, influence the dominant microbial community.

HEALTH EFFECTS OF TRACE ELEMENTS AND METALS IN EARTH MATERIALS

Both toxic and beneficial trace elements are naturally present in soils as a consequence of soil parent minerals and as a result of atmospheric deposition of natural materials (e.g., volcanic ash). They are also present as a result of anthropogenic inputs, including application of treated sew-

TABLE 5.1 Trace Element Concentrations (mg kg^{-1} dry weight) in Agricultural Soils and Food Crops

Element	Common Range for Agricultural Soils	Selected Average for Soils	Typical Range for Food Crops
Arsenic (As)	<1–95	5.8	0.009–1.5
Barium	19–2368	500	1–198
Boron	1–467	9.5–85	1.3–16
Cadmium	0.01–2.5	0.06–1.1	0.13–0.28
Cobalt	0.1–70	7.9	8–100
Chromium	1.4 - 1300	54	0.013–4.2
Copper	1–205	13–24	1–10
Fluorine	10–1360	329	0.2–28.3
Mercury	0.05–0.3	0.03	0.0026–0.086
Molybdenum	0.013–17	1.8	0.07–1.75
Nickel	0.2–450	20	0.3–3.8
Lead	3–189	32	0.05–3.0
Selenium	0.005–3.5	0.33	0.001–18
Silver	0.03–0.9	0.05	0.03–2.9
Tin	1–11	—	0.2–7.9
Vanadium	18–115	58	0.5–280
Zinc	17–125	64	1.2–73

SOURCE: Alloway (2005).

age sludge and fertilizers and atmospheric deposition from industrial sources. Because of differences in the mineralogy of the parent materials and the variable levels and broad range of contamination from anthropogenic sources, soils are found with a wide range of trace metal concentrations. Trace element concentrations in agricultural soils can vary by two to three orders of magnitude (see Table 5.1).

Chaney (1983) classified trace elements in agricultural soils that received sewage sludge and other wastes according to their potential for risk. At a soil pH of 6-8, the low solubilities or strong adsorptions of silver (Ag), gold (Au), chromium (Cr), fluorine (F), galium (Ga), mercury (Hg), lead (Pb), palladium (Pd), platinum (Pt), silicon (Si), tin (Sn), titanium (Ti), and zirconium (Zr) essentially preclude significantly increased concentrations in plants even when the soils are greatly enriched in these elements. Increased concentrations of aluminum (Al), arsenic (As), boron (B), barium (Ba), beryllium (Be), copper (Cu), iron (Fe), manganese (Mn), nickel (Ni), vanadium (V), and zinc (Zn) in plants are insufficient to adversely affect animals because the element causes phytotoxicity, the element is well tolerated by animals, and/or the maximum increased level in plants is lower than the toxic level to animals. Elements that are easily translocated within the plant and can reach foliar levels sufficient to cause adverse health effects include cadmium (Cd), cobalt (Co), molybdenum (Mo), and selenium (Se).

Metal Partitioning in Soils

The bulk composition of a soil is rarely a good predictor of the availability of elements or relative risk. Rather, the bioavailability or the amount of the metal in soil solution is more important. Thus, partitioning of trace elements between soil and the soil solution governs their mobility and availability for uptake by plants and other organisms (Allen, 2002). The uptake of trace elements by earthworms is principally from soil solution, rather than from food or ingested soil particles (Saxe et al., 2001). It is therefore important to understand the partitioning of metals between soil and soil solution in order to assess the potential for ecotoxicological effects and—because many trace elements are required nutrients—to assess the nutrient status of soils.

A number of recent reviews of the processes controlling chemical partitioning show that a variety of factors influence metal mobility in soils, with the most important factor being soil pH (Adriano, 2001; Kabata-Pendias and Pendias, 2001; Allen, 2002; Alloway, 2005). Sorption of cationic trace elements increases with increasing pH, generally with a sharp increase occurring over about a 2 pH unit range. Conversely, elements that are anionic, including oxyanions such as arsenate and chromate, are more strongly bound at lower pH values. The permanently charged surface sites of clay minerals sorb metals by ion exchange or chemisorption onto the variably charged surfaces of metal oxides and hydroxides as well as those of amorphous aluminosilicates (Stumm, 1990). The organic matter in soils also participates in the partitioning reactions by forming stable complexes with many metals, and in many cases binding of metals by organic matter is the most important process (Lee et al., 1996; Tipping, 2002).

Oxidation reduction conditions directly influence element mobility in soils, potentially altering both dissolved and surface-bound species. The classic example is chromium, which occurs as trivalent Cr^{+3} under reducing conditions and as hexavalent Cr^{+6} under oxidizing conditions. Whereas Cr^{+3} has limited mobility due to formation of the sparingly soluble chromium hydroxide, hexavalent chromium is much more mobile than trivalent chromium and is a known carcinogen. Another example is arsenic, which can exist as arsenate (As^{+5}) or arsenite (As^{+3}). Arsenite is more mobile and more toxic. Under oxidizing conditions, iron and manganese form insoluble oxides and hydroxides that can scavenge and sequester a variety of trace elements or provide a suitable surface for their sorption (Stumm and Morgan, 1996). However, under reducing condi-

BOX 5.2
Use of Geochemical Models for Public Health

A number of models are available to characterize the equilibrium speciation of soil solutions, and these models can also provide data for physiological studies of metal toxicity. Among the commonly used models in such geochemical studies are PHREEQC (Parkhurst and Appelo, 1999), MINEQL+ (Schecher and McAvoy, 1998), and MINTEQA2 (EPA, 2006b). These models are capable of modeling solution and solid-phase speciation, including organic and inorganic complexes of metals, and they can provide solution speciation and saturation indices (whether a mineral should dissolve or precipitate) for a broad range of minerals. Sorption can be included by a variety of processes, ranging from simple Langmuir isotherms to various surface complexation models. However, while these models are adequate to evaluate metal complexation by simple ligands, they are not yet capable of considering reactions between metals and complex humic substances with broad ranges of binding energies.

Two models represent the state of the art for simulating interactions between humic materials and metals—SWAMP (Sediment Water Algorithm for Metal Partitioning; Radovanovic and Koelmans, 1998) and SCAMP (Surface Chemistry Assemblage Model for Particles; Lofts and Tipping, 1998). Both have been used to characterize the interactions between dissolved metals and suspended particulate matter. The SWAMP model couples a speciation model with a surface complexation model, and expresses the stability constant (K_d) for suspended solids in which metals complex with inorganic species, organic species, and solid surfaces. SCAMP uses a similar chemical equilibrium method, with the interactions between humics and metals described using "Model V" (Tipping, 1994, 1998, 2002).

tions, metal-respiring bacteria can reduce both Fe and Mn oxy/hydroxides to much more soluble Fe^{+2} and Mn^{+2} states (Lovley, 1987). The reduction of these oxides releases incorporated complexed elements into solution, together with any surface-bound elements (which may be nutrients or contaminants). The kinetics of the sorption/desorption reaction are complex and poorly understood. Elements added to soil, especially metals, slowly become more stable over time and less likely to partition into the soil solution. The metal hydroxide precipitates mentioned above account for some of this behavior at higher pH, but other processes may also be involved, both at the higher pH values at which metal hydroxides and mixed metal-aluminum lattice double hydroxides form and at the lower pH values at which these precipitation processes would not be effective (Stumm and Morgan, 1996) (see Box 5.2).

Because plants ultimately derive their trace metals from soil solutions rather than directly from soil minerals, total element concentrations in soils are often poor predictors of the bioaccessibility and bioavailability of trace elements. Increasing the solution concentration of a trace element by complexation (for instance by chloride or sulfate), by lowering pH for cationic metals or by raising the pH by adding lime (CaO) for anionic species, increases the solubility and uptake of the element into the root. In contrast, hyperaccumulators—such as some *brassica* species—are capable of accumulating greater than 1% of specific metals, and these plants have been used for remediation of metal-contaminated sites. It seems probable that other plant species may accumulate significant metal concentrations and that these could serve as a food source for humans. More research is needed to develop analytical and modeling methods to better describe the bioavailability and geoavailability of elements in soils.

Health Effects of Bioaccumulation of Trace Toxic Metals

It has become increasingly clear that simple, single-element models are often inadequate to explain disease associated with nutrition, element toxicity, or element deficiency associated with foods. Cadmium, for example, stimulates the growth of malignant prostate cells in vitro, whereas selenium, at a critical concentration, inhibits this growth (Webber, 1985). Zinc is considered an essential nutrient for normal prostate growth. Although the precise role of these elements is not clear, biopsies of normal and abnormal prostates have confirmed that cadmium and zinc concentrations are higher in prostates with tumors (Brys et al., 1998; Ogunlewe and Osegbe, 1989). Improved understanding of the synergistic and/or antagonistic trace element interactions in dietary constituents is needed. Importantly, trace element interactions also occur in soil, and these interactions influence the amount of each element taken up by plants and subsequently consumed by humans.

Cadmium (Cd)

The human body burden of cadmium has increased over the past 100 years due to an increase in environmental and industrial pollutants (Thrush, 2000), leading to a range of health effects (see Box 5.3). Individual body burden can increase with poor diet and nutritional status (e.g., as a result of vitamin C and zinc deficiency). In the geological environment, cadmium usually occurs in minerals combined with other elements such as oxygen, chlorine, or sulfur. Cadmium is widely used in industry, where it is found in batteries, pigments, plastics, and metal coatings. It also enters the environment naturally from weathering and from the mining and

BOX 5.3
Human Health Effects of Excess Cadmium in Soil

Cadmium is a nonessential trace element that has been identified as the source of a number of human health problems. The exposure pathway is generally from foodstuffs grown on soil containing elevated levels of cadmium, principally as a result of emissions from mining and smelting of ores and from the application of sewage sludge and phosphatic fertilizers to agricultural land. Additionally, smokers are exposed due to the presence of cadmium in tobacco.

Cadmium-induced disease in humans, occurring principally as a consequence of long-term consumption of cadmium-contaminated rice, is manifested as proximal tubular renal dysfunction. The most severe consequences of cadmium contamination occurred in the Jinzu Valley in Toyama Province, Japan, where mining and smelting operations prior to World War II resulted in contamination of the rice paddy soils with cadmium, lead, and zinc (Alloway, 2005). The flooding and drying out of the paddy fields caused changes in chemical speciation, particularly that of cadmium. Cadmium is immobilized as CdS under flooded and reduced conditions, but under oxidizing conditions it becomes released as Cd^{+2}, which is available to be taken up by the rice plant and translocated to the grain. Rice in the Jinzu Valley was significantly elevated in cadmium content—the average concentration of cadmium in rice grown on contaminated paddy soils was 0.7 mg kg^{-1}, more than 10-fold greater than that in local uncontaminated rice samples. The mean cadmium intake for residents of the Jinzu Valley was approximately 600 mg per day, which is about 10 times the maximum tolerable intake. The most severely affected were women who had several children, who suffered kidney damage and a skeletal disorder know as *itai-itai* (or "ouch-ouch" when translated to English) because of the pain suffered when their bodies were touched. More than 200 women were disabled by the disease and another 65 died from its effects.

Although the concentration of cadmium in food has often been considered the predominant factor to be considered for body burden, numerous other factors are also relevant. In particular, the nutrient status of an individual with respect to zinc, iron, and/or calcium can have a profound effect on the rate of cadmium absorption from the gut (Reeves and Chaney, 2002). Nutrient status, and not solely the concentration of cadmium in rice, must be considered when assessing the risk of dietary cadmium exposure.

processing of rocks and minerals (CDC, 2005). Environmental cadmium pollution occurs in many parts of the world through a combination of land contamination (through fertilizers and sludge application) and water contamination (through irrigation and industry), resulting in cadmium introduction into the food chain. Excess consumption of lamb, kidney, alcohol, grains, and oysters can increase the body burden, and industry-related activities can provide direct occupational exposure (Thrush, 2000). Trace element interactions also occur in soils, ultimately affecting the bioavailability of elements and their subsequent plant uptake. Specifically, high bioavailable concentrations of zinc can reduce the amount of cadmium taken up by plants (Cataldo and Wildung, 1978).

Selenium as a "Protective Factor"

Chemoprevention is the administration of agents to prevent the development of cancer (Platz and Helzlsouer, 2001). Chemoprevention of prostate cancer can be assisted by antioxidants to combat oxidative stress and by the inhibition of androgenic stimulation either pharmacologically or by modification of lifestyle factors such as diet. Selenium is an essential trace element found in varying concentrations in the soil and as organic complexes in foods such as meats, eggs, dairy products, bread, and seafood. Selenium levels in food are largely dependent on the soil content in the region where the food is grown (Combs and Combs, 1984) and therefore intakes vary geographically. Populations living in parts of the world with low-selenium soils who depend on domestic food production may ingest very little selenium and could be at risk of selenium deficiency (Vogt et al., 2003). Many ecological studies have established an inverse correlation between soil selenium levels, prostate cancer mortality, and mortality from other cancers (Clark et al., 1991; Fleet, 1997; Shamberger and Willis, 1969). One study found that men taking selenium supplements for five years had a 65% reduction in the incidence of prostate cancer (Clark et al., 1996). However, another large study found no association between baseline selenium and prostate cancer during nine years of follow-up monitoring (Hartman et al., 1998). Platz and Helzlsouer (2001) suggested that the difference between the findings of Clark et al. (1996) and Hartman et al. (1998) may be due to a difference in actual selenium exposure—one factor might be that the latter study was carried out in Finland, known to have low selenium levels in soil but where fertilizer fortified with selenium had just been introduced. Inverse associations between other cancers and levels of environmental or blood selenium have been recorded (Rayman, 2005).

Although clinical studies have focused on selenium supplementation as a protective factor in reducing prostate cancer incidence (Brawley and

Barnes, 2001; Nelson et al., 1999), the effect of low-selenium bioavailability on the risk of prostate cancer or benign prostatic hyperplasia (BPH) has not been addressed. With the availability of data from community-based studies on the natural history of BPH and placebo-controlled clinical trials, interest is shifting beyond short-term effects on symptoms to reducing the risk of long-term negative outcomes and BPH progression (Roehrborn, 2000).

Zinc as a "Protective Factor"

Zinc is a homeostatically regulated essential mineral present in red meat, poultry, grains, dairy, legumes, and vegetables. It is a critical soil nutrient, and deficiency of zinc in soil can impact crop yield and the nutritive quality of the resulting food crop (Adriano, 2001). Human zinc deficiency has also been associated with geophagia, where the ingestion of soils rich in zinc actually decreased zinc absorption (Hooda et al., 2004). Zinc is a component of numerous metalloenzymes and is important for cell growth and replication, osteogenesis, and immunity. Zinc may also act as an antioxidant by stabilizing membranes in some cell types.

The normal human prostate accumulates the highest zinc levels of any soft tissue in the body—10 times higher than for other soft tissue (Costello and Franklin, 1998). Zinc levels in prostate cancer cells are markedly decreased compared with nonprostate tissues, and there is evidence that zinc inhibits human prostate cancer cell growth (Liang et al., 1999). Cancer cells from prostate tumors have been found to lose their ability to amass zinc (Costello and Franklin, 1998).

Reduced red meat consumption and increased cereals in the diet may reduce the intake and bioavailability of zinc (Gibson et al., 2001). Both dietary and biochemical data suggest that the current Western diets of the elderly may result in a risk of zinc deficiency.

Arsenic

The distribution of naturally occurring arsenic and the health effects of arsenic exposure have been reviewed in several recent review articles (Oremland and Stolz, 2003; Smedley and Kinniburgh, 2005; Centeno et al., 2005) and in Chapter 3 above. Here the microbial role in determining the speciation and bioavailable concentrations of arsenic in soils and the resultant effects from arsenic ingestion through food are described (see Box 5.4).

Soil microorganisms can transform and metabolize arsenic species found in soil, both as a pathway to conserve energy and to provide a defense mechanism against the toxic effects of arsenic. Some soil microbes can use arsenate as a terminal electron acceptor to reduce As^{+5} to As^{+3} (Jackson and Dugas; 2003), conserving energy from the oxidation of organic carbon in anaerobic environments but producing the more toxic form of arsenic. Other microbes (or even the same organism, e.g., *Thermus* HR-1: see Gihring and Banfield, 2001) oxidize arsenite to arsenate, sometimes using arsenite as substrate and conserving energy as a chemoautotroph (e.g., Oremland et al., 2002). Oxidized arsenic (arsenate) is accidentally taken up by micro-organisms as part of the phosphate transport system, due to the similarity of the As^{+5} oxyanion species to inorganic ortho-phosphate, and the effects of arsenic toxicity can be increased or decreased by pH, temperature, and coexposure to other metals. A variety of bacteria have developed resistance to extreme arsenic concentrations, reducing arsenate to arsenite intracellularly and pumping out arsenite (Silver and Keach, 1982). The microbial response to toxic arsenic is largely to change it to the most toxic and mobile species, which are then available to be taken up in crops or infiltrated to groundwater. Both As^{+5} and As^{+3} are taken up by rice (Abedin et al., 2002) and vegetables (Queirolo et al., 2002).

OPPORTUNITIES FOR RESEARCH COLLABORATION

Interdisciplinary collaboration will be essential to advance our understanding of the complex interrelationships at the intersection of agriculture, soils, microbiology, and public health. Conceptually, the soil environment controls the variety and quantity of elements and nutrients taken up by plants and therefore the elemental composition of plants and their nutritional status. Ultimately, this manifests itself in terms of what is eaten by humans, and therefore biogeochemical cycling in soils strongly impacts what people ingest. Soil, the easily disturbed interface between humans and the geological substrate, constitutes a ripe area of research for the earth science and public health communities. High-priority collaborative research activities are:

1. To determine the influence of biogeochemical cycling of trace elements in soils as it relates to low-dose chronic exposure via toxic elements in foods and ultimately its influence on human health. For example, it is well known that zinc and cadmium compete for plant uptake in soils and that zinc protects against excess cadmium uptake. Similar protective mechanisms influence the bioavailability of cadmium in the human body. However, in general, little is known about these elemental interactions and the influence of mixtures of elements on bioavailability in both soils and the human body. Similarly, little is known about low-dose chronic exposure via toxic elements in foods.

BOX 5.4
Arsenic-Contaminated Food

The contribution of arsenic in food to total human arsenic intake has not been extensively studied, but there is evidence that water is not the sole source of this toxic element. Fish and shellfish are a recognized source of total arsenic, and while these sources may be particularly high in organo-arsenicals, this represents the least toxic form of arsenic. The contribution of food crops to total arsenic intake, particularly the more toxic inorganic forms, is poorly understood.

Dietary selenium status has been shown to influence arsenic excretion in animal models. Gregus et al. (1998) noted that selenium facilitates the excretion of inorganic arsenic metabolites in rats. Selenium supplementation (organic forms of selenium) has been shown to be helpful against poisoning from arsenic and other toxic elements in mice (Andersen and Nielsen, 1994). Similar results have been noted among humans. Recently, Hsueh et al. (2003) found that, in a Taiwanese population exposed to inorganic arsenic via drinking water, urinary arsenic levels significantly increased as urinary selenium levels increased. These observations were recently confirmed in another independent study conducted in Chile from a population exposed to moderate levels of arsenic in their drinking water (~40 $\mu g\ L^{-1}$) (Christian et al., 2006). The results from these studies suggest that, in populations exposed to arsenic, dietary selenium intake may be correlated with urinary arsenic excretion and may alter arsenic methylation.

Arsenic from Coal in China

Domestic coal combustion has had profound adverse effects on the health of millions of people worldwide. In China alone, several hundred million people commonly burn raw coal in unvented stoves, a process that permeates their homes with high levels of toxic metals and organic compounds. At least 3,000 people in Guizhou Province in southwest China, where coal samples contain up to 35,000 ppm arsenic, suffer from severe arsenic poisoning. Although fresh chili peppers contain less than 1 ppm arsenic, Zheng et al. (1996) showed that chili peppers dried over open coal-burning stoves have on average more than 500 ppm arsenic and therefore may be the principal vector for the arsenic poisoning (Figure 5.1). Significant amounts of arsenic may also come from other tainted foods, from dust ingestion (samples of kitchen dust contained as much as 3,000 ppm arsenic), and from inhalation of indoor arsenic-polluted air. In this area, the arsenic content of drinking water samples was below the Environmental Protection Agency's drinking water standard of 10 ppb and does not appear to be an important factor.

Data describing the concentrations and distributions of potentially toxic elements in coal may assist people dependent on local coal sources to

FIGURE 5.1 Chili peppers dried over open, unvented, coal-burning stoves are the main pathway for chronic arsenic exposure in Guizhou Province, China. SOURCE: Finkelman et al. (2001).

avoid those deposits that have high concentrations of toxic metalloids and compounds. Information on the modes of occurrence of potentially toxic elements, and the textural relations on the minerals and macerals in which they occur, may help scientists anticipate the behavior of the potentially toxic compounds and metals during coal use. This type of characterization offers an opportunity for geoscientists and public health professionals to directly contribute to the resolution of a major public health issue.

Arsenic-Contaminated Food in Chile

The Northern Region II of Chile has both natural and anthropogenic sources of environmental arsenic. The major river basin in this region is the Rio Loa, sourced in the highly contaminated El Tatio geyser basin where natural hot spring waters contain 30–50 mg L^{-1} total arsenic. The Rio Loa also receives arsenic from runoff waters and airborne emissions due to extensive copper mining activities in the region.

Crops grown in the village of Chiu Chiu have substantially elevated levels of arsenic and, depending on the arsenic load in the drinking water, food represents 4–25% of the total arsenic intake. The arsenic principally occurs as inorganic As^{+3} and As^{+5} and carrots, for example, preferentially take up the most toxic As^{+3}. While the level of arsenic in the food crops of this region is generally below Chile's regulatory limits, concentrations are significantly higher than would typically be found in foods in other regions and represent a significant component of the total arsenic intake by local inhabitants.

2. To determine the distribution, survival, and transfer of plant and human pathogens through soil with respect to the geological framework. The historical approach for evaluating pathogens in soil has been to describe the soil composition and structure in broad agricultural categories such as "sandy clay loam," while the geological approach to a soil has been either to simplify the microbial biomass and community into the number of colony-forming units, or simply to characterize it all as soil organic matter. Earth scientists who have examined microbial communities are almost always interested in geochemically significant guilds that perform geological functions, rather than the pathogens that are present. Collaboration would involve earth scientists who would be responsible for characterizing the biogeochemical habitat, such as the mineralogy, exchangeable cations, mobile metal species, and/or reactive geochemical surfaces, including sources of nutrients or the presence of antagonistic and/or synergistic metal species. Microbiologists would characterize the microbial community that surrounds the pathogen and examine its viability in different biogeochemical habitats. Public health specialists would examine the incidence of human and plant disease from soil pathogens as a function of the biogeochemical framework, and the role of soils in long-term survival of pathogens and as reservoirs of pathogens. This should also be examined in a bio-security context with respect to food pathogens and food safety. Finally, there is a need to evaluate the potential for plant uptake of human pathogens introduced into soils.

3. To improve understanding of the relationship between disease and metal speciation and between disease and metal-metal interaction. In this research, earth scientists would characterize metal abundance and metal speciation in soils and the mobility and availability of these metals to the biosphere; microbiologists would characterize the microbial populations and mechanisms that are responsible for metal species transitions in soil environments; and public health specialists would use spatial information on the distribution of metal speciation to examine the incidence of specific disease

6

Earth Perturbations and Public Health Impacts

This chapter considers the crosscutting issues associated with perturbations of the earth's environment and the public health consequences of such perturbations. Not only are natural disasters considered, such as volcanic eruptions and earthquakes, but also the public health consequences of anthropogenic perturbations such as those caused by the extractive (natural resources) industries.

PUBLIC HEALTH CONSEQUENCES OF NATURAL DISASTERS

Numerous public health issues are caused by natural disasters—extreme geological and geophysical events (see Table 6.1). Approximately 75% of the world's population lives in areas commonly affected by earthquakes, tropical cyclones, floods, and/or droughts (UNDP, 2004), and these natural events, together with volcanic eruptions, landslides, land subsidence, and coastal inundation, produce profoundly devastating worldwide human health and socio-economic impacts. The United Nations Development Programme report estimates that more than 1.5 million people have died in the past 20 years as a result of natural disasters, mostly in Asia and around the Pacific Rim. Natural disasters can increase the incidence of communicable disease among displaced communities and cause profoundly negative sociological effects.

The health consequences of disasters may be separated into two components—immediate (or direct) and longer term (or indirect). Both are exacerbated by the fact that disasters frequently destroy or damage local

TABLE 6.1 Fatalities (rounded to the nearest hundred) from Selected Natural Hazards, 1960–1987, and the Largest Single Disaster for Each Hazard

Hazard Type	Deaths	Largest Single Event and Year	Deaths
Coastal inundation	761,400	Eastern Pakistan (Bangladesh), 1970	500,000
Earthquakes	557,900	Tangshan, China, 1976	250,000
River floods	40,100	Vietnam, 1964	8,000
Landslides, mudflows	39,600	Peru, 1970	25,000
Volcanic eruptions	27,500	Columbia, 1985	23,000
Tornados	4,500	Eastern Pakistan (Bangladesh), 1969	500

NOTE: More recent seismic events include the 2004 Sumatran tsunami (more than 285,000 fatalities) and the 2005 northeastern Pakistan earthquake (83,000 fatalities).
SOURCES: Bryant (1991), Munich Re Group (2000).

medical care facilities and the local public health infrastructure and disrupt and destroy transportation systems, communications facilities, and social services. Food and water supplies may also be destroyed, and even where they are not, disruption of the transportation system may make it difficult to ship adequate food supplies to affected regions, resulting in poor nutritional status and intensifying disease outbreaks. This impedes disaster recovery, and medical treatment during the acute phase of a disaster can be extremely difficult. In many cases, morbidity and mortality from the long-term health consequences may exceed the deaths resulting directly from the disaster (UNDP, 2004).

Although there has been a long tradition of addressing the human responses to natural disasters and hazards at governmental, institutional, and behavioral levels (e.g., Burton et al., 1978; Hewitt, 1997; Platt, 1999; Smith, 2001), far less attention has been paid to the public health consequences of disasters, particularly within what is conventionally considered the "natural hazards" literature in the social and behavioral sciences (Mileti, 1999). Recently, increased attention has been devoted to health issues associated with natural disasters—these include direct mortality from trauma, indirect mortality and morbidity from infectious diseases, and mental health problems such as post-traumatic stress disorder (Benin, 1985; Noji, 1997, 2005; Mileti, 1999).

The public health impacts of natural disasters have resulted in the development of the field of "disaster epidemiology" (e.g., Wasley, 1995).

Epidemiological activities during a disaster include disease control and surveillance as well as injury epidemiology. Geographic Information System (GIS) information is now routinely used, ideally to detect disease clusters in real time. Prior to a disaster, epidemiological information is indispensable for the identification of vulnerable populations, with the use of mathematical and statistical risk models to identify areas that are vulnerable to natural disaster impacts, providing the basis for GIS map production (see Chapter 7). The U.S. Geological Survey already maintains a series of maps that depict earthquake risk as part of the National Seismic Hazard Mapping Project.[1] These maps are available down to the zip code level throughout the United States. Population density maps can be overlaid on hazard maps to analyze the spatial concurrence of earthquake risk and population location. Many other countries with high levels of seismic risk have similar programs. Virtually any earthquake that has caused significant damage to human habitat has represented a potential public health problem because of the coincidence of risk and population distribution.

Vulnerability to both short- and long-term health effects is greatest among the impoverished and in the poorer parts of the world. One estimate is that the 66% of the world's population living in the poorest countries accounts for 95% of the mortality due to disasters (Anderson, 1991). The immediate consequences are usually injuries or deaths due to trauma. Earthquakes cause trauma due to the collapse of structures and other edifices, and can also cause coastal flooding due to tsunamis. The recent 2004 Sumatran earthquake and tsunami caused drowning, traumatic injury, and structural collapse over a huge area around the Indian Ocean. A recent case control study in Taipei, Taiwan, suggests that socioeconomic status, preexisting health status, physical disability, and location were major predictors of mortality in the 1999 Chi-Chi earthquake (Chou et al., 2004).

Mitigation of the adverse impacts of future hazards necessitates their clear recognition, prediction, and early warning; quantification of the processes involved; accurate assessment of associated risks; and hazard avoidance or technological mitigation to reduce vulnerability. The first three steps require scientific and engineering investigations as well as an understanding of the social and population dynamics associated with hazards. The fourth component involves preventative measures such as redesigning and reinforcing buildings, bridges, and dams, and constructing all-weather shelters, dikes, and seawalls. This may also involve strength-

[1]See *http://eqhazmaps.usgs.gov/*.

ening the construction of houses, changing building codes, and improving emergency response systems and public health infrastructure. Broad public understanding of the dangers posed by natural hazards is absolutely crucial for hazard avoidance and the technical mitigation process, and collaboration between the earth science and public health communities is a critical component for increasing public knowledge.

Infectious Disease Impacts

Longer term public health threats from natural disasters include infectious diseases, often vectorborne. For example, following flood inundations in tropical areas (either from storms or tsunamis), ecological conditions are frequently optimal for anopheline mosquito reproduction if they are already present in the ecosystem, resulting in an increase in malaria in vulnerable populations (NRC, 2002c; Toole, 1997). In 1963, 75,000 cases of *Plasmodium falciparum*—a potentially deadly form of malaria—were recorded in Haiti following Hurricane Flora.

Many disasters result in population displacement and migration, frequently to refugee camps with high population densities, and in such environments infectious diseases result from both overcrowding and poor sanitation. Ecological and social conditions are conducive to the spread of enteric diseases that include cholera (Kalipeni and Oppong, 1998). Refugee camps are also associated with the spread of diseases via the respiratory route, including meningitis, tuberculosis, and multiple drug-resistant tuberculosis (Rutta et al., 2001), as well as HIV/AIDS and other sexually transmitted infections (UNAIDS, 2005; Salama and Dondero, 2001). In recent years, vulnerability to HIV/AIDS in sub-Saharan Africa has increased as a result of drought and famine; in turn, the famine has been exacerbated by the loss of agricultural workers who have succumbed to AIDS (UNDP, 2003, 2004).

Noninfectious Disease Impacts

Longer term health effects of disasters are not limited to infectious diseases. Food production may decline due to the agricultural effects of a natural disaster. For example, following the 2004 Indian Ocean tsunami, the incursion of saltwater in parts of Sri Lanka through mangrove swamps into inland rice paddies led to increased salinity. This will probably lead to decreased productivity (IWMI, 2005). Malaysia faces a similar problem, and both countries are seeking saline-tolerant forms of rice to mitigate the impacts.

A growing body of research suggests that there is a notable increase in acute myocardial infarctions (heart attacks) following earthquakes, up to five times the normal risk (Tsai et al., 2004; Ogawa et al., 2000); putatively, this is due to the extreme stress and fear caused by severe earthquakes. Increased myocardial infarction cases were recorded following the Northridge earthquake in 1994 and the Hanshin (Kobe) earthquake of 1995. The Northridge earthquake also triggered an excess number of out-of-hospital cardiac arrests. Most of these cardiac arrests were due to underlying atherosclerosis, suggesting that the earthquake was a triggering event for deaths that would probably have occurred in the near future (Leor et al., 1996).

Psychosocial Stress Impacts

Disasters have also been linked to psychiatric disorders, most notably to post-traumatic stress disorder (PTSD). This is not surprising, as the etiology of PTSD is usually some sudden, extremely stressful, emotionally disruptive and wrenching event, frequently involving the death of others and the threat of death to oneself. There was evidence of PTSD in 68% of 160 disasters that were sampled in one review of natural disasters occurring between 1981 and 2001 (Norris et al., 2002). The severity was greater in developing countries than in developed countries. In the Mexican floods of 1999 the prevalence of PTSD was a striking 46% in Tezuitlan, and there was significant comorbidity with depressive disorder directly attributable to the personal and property losses associated with the floods (Norris et al., 2004). Similar comorbidity was noted in Turkey following the 2003 Bingol earthquake (Ozen and Sir, 2004). In addition, those affected by the 1999 Chi-Chi earthquake in Taiwan, perhaps compounded by the overall economic stress in Asia, were 1.46 times more likely to commit suicide after the earthquake (Chou et al., 2003). Mental health needs following disasters are significant and are not as well addressed as are the "physical" health needs (although admittedly the biological bases of psychiatric disorders militate against a dichotomy between "mental" and "physical").

LAND COVER CHANGE AND VECTORBORNE DISEASES

Human-induced land surface changes are the primary drivers of a range of infectious disease outbreaks and also modifiers of the transmission of endemic infections (Patz et al., 2000). Such anthropogenic landsurface changes include (1) deforestation and road construction; (2) agricultural encroachment and water projects (e.g., dam building, irrigation, and wetlands modification); (3) urban sprawl; and (4) extractive industries such as mining, quarrying, and oil drilling. These land surface changes cause a cascade of factors that heighten health threats, including infectious disease emergence, forest fragmentation, pathogen introduction, pollution, poverty, and human migration. Natural geological determinants of disease primarily relate to the larval stage of vectorborne diseases, when soils and surface water availability factor into insect breeding site availability and quality. These are important but complex issues that are only understood for a few diseases. For example, recent research has shown that forest fragmentation, urban sprawl, and biodiversity loss are linked to increased Lyme disease risk in the northeastern United States (Ostfeld and Keesing, 2000; Schmidt and Ostfeld, 2001). Expansion and changes in agricultural practices are intimately associated with the emergence of Nipah virus in Malaysia (Chua et al., 1999; Lam and Chua, 2002), *Cryptosporidiosis* in Europe and North America, and a range of foodborne illnesses globally (Rose et al., 2001).

Rates of deforestation have grown exponentially since the beginning of the twentieth century. Driven by rapidly increasing human populations, large swaths of species-rich tropical and temperate forests, as well as prairies, grasslands, and wetlands, have been converted to species-poor agricultural and ranching areas. In parallel with this habitat destruction, there has been an exponential growth in human-wildlife interaction and conflict. This has resulted in exposure to new pathogens for humans, livestock, and wildlife (Wolfe et al., 2000). Deforestation, and the processes that lead to it, have a number of adverse consequences for ecosystems. Deforestation decreases the overall habitat available for wildlife species. It also modifies the structure of environments, for example, by fragmenting habitats into smaller patches separated by agricultural activities or human populations. Increased "edge effect" (from a patchwork of varied land uses) can further promote interaction among pathogens, vectors, and hosts. This edge effect has been well documented in the case of Lyme disease (Glass et al., 1995). Similarly, increased activity in forest habitats (through behavior or occupation) appears to be a major risk factor for contracting leishmaniasis, a disease caused by protozoa and transmitted by sandflies (Weigle et al., 1993). Evidence is mounting that deforestation and ecosystem changes have serious implications for the distribution of many other microorganisms and the health of human, domestic animal, and wildlife populations.

Landscape epidemiology is based on the concept that geology and climate interact to form a characteristic vegetation cover dictated by the available mineral composition of the soil and substrate together with patterns of temperature and precipitation (Fish, 1996). Vegetation, in turn, provides a microclimate (temperature humidity, shade, etc.) and resources (leaves, fruits, nectar, etc.) which determine the species composition and abundance of vertebrates and vectors, which in turn support the natural transmission of specific pathogens. This generalized model is applicable

BOX 6.1
Lyme Disease in the United States

First discovered in the early 1970s and described as an epidemic of juvenile arthritis confined to the coastal community of Lyme, Connecticut (Steere et al., 1977), Lyme disease is now known to be endemic in 19 states (Nadelman and Wormser, 2005). The force behind this epidemic has been environmental change, which has increased contact between humans and a spirochete bacteria (*Borrelia burgdorferi*) transmitted by a tick (*Ixodes scapularis*). Adult ticks feed primarily on white-tailed deer, and new populations of ticks have spread rapidly over the northeastern United States through deer movement. This was assisted by reforestation prompted by farm abandonment in many areas of the northeastern United States during the early twentieth century. Combined with a marked expansion of the human population, this sequence of events has resulted in wider exposure to natural tickborne pathogens, including *B. burgdorferi* (Falco et al., 1995). This has resulted in a continuous increase of Lyme disease cases reported to the Centers for Disease Control and Prevention (CDC) over the past 20 years, despite enormous efforts to educate the public on prevention measures and the development of a Lyme disease vaccine. Geological factors, particularly sedimentary bedrock and soil type, play a key role in determining tick habitat and the risk of acquiring Lyme disease (Guerra et al., 2002).

to all pathogens of nonhuman origin (zoonoses), whether vectorborne or directly transmissible. Humans are not required for their maintenance in nature, although zoonotic pathogens may ultimately adapt to direct human-to-human transmission modes.

Microbial agents causing infectious diseases in humans often originate from processes and events occurring in the natural environment. Most of the so-called emerging diseases are caused by infectious agents of wildlife that have either recently adapted to infect humans or are pre-adapted and have recently come into opportunistic contact with humans (Taylor et al., 2001). These include some of the most important pathogenic agents that have caused major epidemics in humans, such as HIV/AIDS, influenza A, Ebola, West Nile virus, and Lyme disease. The current epidemic of Lyme disease in the United States (see Box 6.1) provides a relevant example of how environmental change can result in epidemic disease in humans. The geographic distribution of the risk of Lyme disease for humans can be predicted based on vegetation and climate characteristics that determine the distribution of the vector, wildlife hosts, and the pathogen (Guerra et al., 2002; Brownstein et al., 2003). The extent of human exposure to infected ticks in the environment constitutes risk, which determines the distribution and prevalence of Lyme disease (Fish, 1995).

Application of landscape epidemiology to Lyme disease has had a significant impact on our understanding of this epidemic and on the implementation of prevention measures. Early field studies identified the importance of peridomestic risk in the suburban landscape in the northeastern states (Falco and Fish, 1988a, 1988b), which enabled public health agencies to target high-risk populations for education on preventive measures. The CDC Advisory Committee on Immunization Practices issued guidelines for vaccination against Lyme disease based entirely on a national Lyme disease risk map generated from ecological data on the distribution and prevalence of infection in vector ticks throughout the United States (Fish and Howard, 1999).

The threat of emerging diseases from wildlife and vectors is a continuous, dynamic process, which most likely will accelerate due to human population growth and more extensive environmental change. Therefore, it is imperative that more emphasis be placed on environmental studies of infectious agents in order to balance our overreliance on diagnostics, therapeutics, and vaccines for humans, which at present dominate the biomedical research effort in emerging infectious diseases (Morens et al., 2004).

HEALTH EFFECTS OF RESOURCE EXTRACTION AND PROCESSING

Terrestrial mineral resources include abundant metals, scarce metals, water, soil, building materials, and a wide range of chemicals, including carbon-based fuels and nuclear energy sources. The quarrying of construction materials, the mining of ore deposits and coal, and the drilling for and production of oil and natural gas take place in relatively restricted areas, especially compared with the widespread land surface modifications wrought by agricultural, forestry, urban, and industrial development and the resultant degradation of the air, land, and water environments (Gleick, 1998; Harrison and Pearce, 2000; IPCC, 2001a, 2001b; Wolman, 2002). Over the period of intense resource exploitation activity between 1930 and 1980, less than 1% of the total land area of the United States was directly impacted by coal mining, mineral mining, and petroleum production activities (Johnson and Paone, 1982). However, topographic alteration, groundwater and surface water contamination, and hydrocarbon pollution are far more serious than this small areal percentage might suggest because of the high toxicity of a proportion of the mining waste and petroleum products. The extraction, beneficiation, and usage of earth materials, including fossil fuels, result in deleterious side effects that include environmental degradation and diminished viability of the biosphere in general and human health in particular. As noted in earlier chapters, such health hazards include airborne dusts and gases, soluble chemical pollutants in both surface water and groundwater, and toxic substances in soils, crops, livestock, and manufactured products.

Modern societies are maintained by the extraction of energy, water, and other earth materials far beyond natural renewal rates, providing limits to future human use of such natural resources. As more intensive usage of earth commodities and energy takes place due to increasing global population, the attendant global demand for a better standard of living will result in an increase in the adverse impacts of resource-related health hazards unless steps are taken to address and ameliorate them (e.g., McMichael, 2002). The biological carrying capacity of the earth is finite—hence humanity eventually must reach a managed steady state with the available terrestrial resources and the life support system provided by the biosphere. Reflecting the intense desires of developing nations for an improved standard of living, our own security dictates the need for a much more equitable consumption of mineral commodities and distribution of wealth. However, to achieve mineral resource sustainability (NRC, 1996, 1999e) global research in science and technology must be increased in order to reach much more efficient levels in the development—and especially the conservation of—earth resources (WCED, 1987; Chesworth, 2002; Doran and Sims, 2002). An overridingly important part of this challenge will be to preserve an intact, healthy, functioning biosphere.

Mineral Exploration, Extraction, and Processing

The process of mining ore deposits and coal contains several steps that can potentially expose humans to toxic materials (e.g., Box 6.2). Although the steps can vary for different types of materials, they generally include some combination of extraction, processing and refining, use, and waste disposal. Although modern extraction technologies are much more efficient, and in many countries more highly regulated, than in the past, retrieval of materials from the earth for human use is one of the most serious sources for contamination of soils, water, and the biosphere (Selinus et al., 2005). In the case of metalliferous ores, the greatest environmental contamination generally comes from the mineral processing that occurs after extraction from the mine (e.g., CDC, 2005). This processing produces mine tailings, often consisting of very fine dust that can contain residual amounts of mineral ore and other harmful trace elements. Frequently left open to the environment, the tailings are subject to transport by both wind and water, resulting in contamination of surrounding soils, surface water, and groundwater.

BOX 6.2
Copper and Lead

A number of geological regions—the Lake Superior mineral province, Arizona, northern Nevada–eastern Oregon, southwestern Montana, Indonesia, and northern Chile—contain abundant copper deposits where copper is released to the environment either naturally or as a result of mining and smelting. In aqueous solution, this metal becomes bioavailable as the monovalent cation Cu^{+} in anoxic water and as the divalent cation Cu^{2+} in oxygenated water (Robbins and Harthill, 2003). Copper is not generally concentrated to dangerous levels in the human body, as excess amounts normally are not retained, but can cause diarrhea and stomach disorders. Where excessive amounts are present, toxicity is manifested as chronic pulmonary or liver damage.

Geochemical culminations of lead are commonly associated with zinc in "Mississippi Valley–type" ore deposits. In these occurrences, sulfides of these metals are precipitated from relatively low-temperature hydrothermal solutions passing through, and partly replacing, limestone and dolomite. Other occurrences of lead and zinc sulfides are disseminated in granitic intrusive rocks. Anthropogenic utilization of lead began in earnest with the Romans, but accelerated in the twentieth century with the widespread application as white pigment in paint and as lead additive to gasoline (Mielke et al., 2003; Filippelli et al., 2005). Both usages are now phased out in the United States. However, geochemical concentrations of lead in stream sediments seem to reflect both natural occurrences in mining districts and widespread past utilization by humans. Lead poisoning results in neurological impairment (learning disorders, mental retardation, attention deficit disorder), deafness, cardiovascular disease, and impaired physical development.

In coal mining, the steps of extraction, combustion, and waste disposal can all pose a threat to human health. On the extraction side, mining operations can generate significant amounts of respirable airborne coal dust. Inhalation of this dust can lead to coal workers' pneumoconiosis, a disabling and potentially fatal lung disease. Depending on the chemical composition of coal, its combustion can contribute significantly to atmospheric deposition of trace metals such as arsenic, cadmium, copper, fluorine, mercury, nickel, and selenium. The residues from coal combustion, including fly ash, bottom ash, boiler slag, and flue gas desulfurization sludge, pose serious disposal problems (NRC, 2006b). Burning coal concentrates potentially harmful metals and metalloids—including arsenic, cadmium, chromium, and lead—in the residues. Depending on the disposal methods, combustion residues can contaminate drinking water supplies or groundwater at levels dangerous to human health, and residue transport and handling can also produce airborne particulate matter that poses an inhalation risk beyond the mine area (NRC, 2006b).

Petroleum Exploration, Drilling, and Extraction

In areas that have active or historical oil and/or gas development, there are a variety of environmental impacts that directly or indirectly impact human health. These include waste materials and pollutants generated during drilling and production as well as leakage and inadvertent spills during later parts of the petroleum life cycle.

Drilling and production result in the discharge of produced waters,[2] drill cuttings, and drilling muds that have the potential for chronic effects on benthic communities, mammals, birds, and humans. Although the petroleum industry is now highly regulated and most of the waste products are recycled on site or disposed of in licensed injection wells, historically this material was abandoned on site in unlined pits or "tanks" (see Figure 6.1) that now require remediation to prevent further groundwater con-

FIGURE 6.1 A tank battery showing produced water discharged to an unlined holding pond in Osage County, Oklahoma.
SOURCE: NETL (2006).

[2]Produced water is the nonhydrocarbon fluid produced from an oil or a gas reservoir during drilling and production. It is often hypersaline and may contain high concentrations of dissolved metals.

tamination. The pits themselves represent a threat to wildlife drawn to the water, especially in otherwise arid regions, and there is now a widespread legacy of saltwater contamination in the shallow aquifers in these regions. Health effects may also occur where wildlife and livestock consume saline water in surface pits. The health effects of chronic exposure to produced water for humans have not been directly studied, although the individual compounds found in the saline fluids have been investigated out of context with the oil and gas industry. Generally, the taste threshold for humans is sufficiently low that little saline water is consumed by accident, and salinization of drinking water is immediately apparent. However, some produced fluids contain naturally occurring radioactive material, and special disposal procedures are required for this fluid (Rajaretnam and Spitz, 2000).

Petroleum contamination of soil is common in oil-producing states in the United States, and many landowners have inherited old collection lines and both low- and high-pressure pipelines where small oil leaks were commonplace. Although the health impacts of benzene, toluene, ethylbenzene, and xzylene[3] (BTEX) have been extensively studied and BTEX pollutant plumes are the subject of ongoing remediation in many parts of the country, the health consequences of contamination of soil and water by the full range of oil components are poorly understood. It is not clear whether health problems are associated with long-term exposure to low concentrations of petroleum in water or air, or if degraded oil poses adverse health effects. The individual components in petroleum in groundwater, particularly the aromatic compounds, are now recognized to rapidly degrade due to the metabolic activity of the native microbial consortium (NRC, 2000a). However, the end product of degradation is not solely carbon dioxide or methane but rather can be a variety of complex, partially degraded, carbon compounds that persist in the water and are transported much farther than the primary BTEX-type compounds. These compounds are generally unresolved by standard analytical techniques and remain largely unidentified, recognized only by the dissolved organic carbon (DOC) plume or after extraction and characterization by advanced analytical techniques (Eganhouse et al., 1993).

OPPORTUNITIES FOR RESEARCH COLLABORATION

The crosscutting issues associated with perturbations of the earth's environment, and the public health consequences of such perturbations,

[3]The BTEX group of volatile organic compounds (VOCs) occurs in crude petroleum and petroleum products (e.g., gasoline)

provide a broad range of research opportunities for both individual investigators and collaborative groups. High-priority collaborative research activities are:

1. Development of improved scenarios for risk-based hazard mitigation. Integrated multidisciplinary approaches to scenario development as the basis for natural hazard mitigation strategies are in their infancy. Any assessment of population vulnerability is dependent on the merging of earth science information describing the spatial distribution of hazards with public health information describing population characteristics and medical response capabilities. Effective scenarios to form the basis of improved response strategies must be scientifically valid and believable for broad acceptance by those charged with disaster response planning. The scientific validation will require collaborative involvement of a broad range of experts from the earth science, public health, emergency management, and engineering communities. Existing tools (e.g., HAZUS for earthquake loss estimation, EPISIM for epidemic simulation, HRAI for hazard risk assessment for public health) provide an indication of the potential for natural hazard mitigation but also emphasize how much is still required in many parts of the nation for the development of improved response capabilities.

2. Analysis of the effect of geomorphic and hydrological land surface alteration on disease ecology, including emergence/resurgence and transmission of disease. Major changes in land use result in profound habitat fragmentation, ecological compartmentalization edge effects, runoff from impervious man-made surfaces, groundwater and/or surface water degradation, and partial or complete loss of habitat—all of which influence disease incidence. The effects of such land surface modifications on the bioaccessibility of disease vectors responsible for disease outbreaks, deleterious chronic disease levels, and human senescence need to be measured individually and collectively in order to better safeguard public health. This will require a seamless integration of geological, hydrological, and epidemiological research efforts in order to prevent, or at least minimize, the adverse effects of land use change on human populations.

Successful collaboration involving earth and health science researchers will allow development of accurate predictive risk assessment models and thus avoidance—or at least amelioration—of a broad spectrum of environmentally related diseases. Because prevention is more effective than medical treatment of diseases, in the long run such cooperative integrated earth science–public health investigations has the potential to save substantial sums of money as well as markedly enhance the quality of life.

Section III

Facilitating Collaborative Research: Mechanisms and Priorities

7

GIScience, Remote Sensing, and Epidemiology: Essential Tools for Collaboration

Geospatial relationships lie at the core of many public health issues, and the integration of remote sensing data, epidemiology, and geographic analysis of disease presents a rich opportunity for collaborative activity at the interface of earth science and public health. Hypotheses generated by the analysis of such geospatial relationships can be tested and refined by analytical and experimental research as the basis for identifying causal relationships. The tools and methods that facilitate analysis include conventional epidemiology, with its subspecialties of genetic, occupational, and environmental epidemiology, as well as remote sensing, Geographic Information Science (GIScience[1]), and the broad field of geospatial analysis that includes spatial statistics and spatial modeling.

GEOSPATIAL ANALYSIS AND EPIDEMIOLOGY

Geographic Information Systems (GISs) used in an epidemiological context are "a simple extension of statistical analyses that join epidemiological, sociological, clinical, and economic data with references to space. A GIS system does not create data but merely relates data using a system of references that describe spatial relationships" (Ricketts, 2003, p. 3).

[1]GIScience refers to the fundamental research principles on which Geographic Information Systems (GIS) are based, incorporating geographic, information, and computer sciences (Goodchild, 1992; NRC, 2006c).

GIScience is the science of collecting, analyzing, and theorizing about geographic information through GISs and geospatial analysis.

There has been increased attention in recent years to the public health applications of GIS, GIScience, and geospatial analysis (e.g., Melnick, 2001; Cromley and McLafferty, 2002; Cromley, 2003). One of the most useful applications of GIS in the public health arena is as a component of exposure and dose assessment. Ideally, exposure to a potential toxin, environmental contaminant, or pollutant would be measured directly at the individual level through monitors that the individual would carry or wear, allowing direct measurement of exposure and subsequent calculation of exposure-response curves. However, such studies are extremely costly, inconvenient, and infrequently carried out for large populations. Instead, exposures may be estimated based on models, and since exposure frequently varies spatially and temporally, a combination of GIS and spatial/temporal models has become an indispensable component of exposure assessment (Nuckols et al., 2004).

A GIS involves the merging of spatially based data—coordinates corresponding to latitude and longitude—with a graphical user interface (GUI). GISs that use data from remote sensing instruments such as aerial photographs and satellite images have become indispensable tools in the development of causal models linking environmental factors and both infectious and noninfectious disease. The spatially referenced database in an epidemiological context usually consists of geocoded (i.e., geo-spatially located) health information, such as the residential locations of people who have contracted a specific cancer, the location of traffic fatalities, or the location of incident cases of myocardial infarction. These data are then superimposed on other data layers, usually geocoded to the same unit as the health data.

Unfortunately, the power of GIS is not always realized in public health applications, or it is misused, because of a lack of understanding of the underlying geographic principles. Just as a person who learns to use statistical software (e.g., STATA, SAS, or SPSS) does not necessarily understand statistics, learning to use GIS software (e.g., ArcGIS, ArcView, MapInfo) does not necessarily ensure an understanding of the underlying principles of geospatial analysis.

CONCEPTS AND COMPONENTS OF GEOSPATIAL ANALYSIS

A GIS is a powerful tool for the analysis of relationships, including causal relationships, between a broad range of measurable variables from the natural sciences—climatic and weather conditions, surface water characteristics, vegetation and land cover, soil geochemistry, and many others—and public health. It is thus a tool and a set of concepts that bring the earth sciences and epidemiology/environmental health into a cause-and-effect relationship with one another. The spatial distribution of Lyme disease can be modeled accurately using GIS and remote sensing at a range of scales to model tick dispersion with reference to a series of environmental variables (e.g., Cortinas et al., 2002; Guerra et al., 2002). The same is true of modeling the effects of climate change on disease distribution, although there is some debate about the accuracy of such models (e.g., Hay et al., 2002; Patz et al., 2002; Tanser et al., 2003; Pascual et al., 2006). Geospatial analysis, or more simply "spatial analysis," uses mathematics and statistics to analyze data patterns that underlie GIS. Many spatial measures and spatial models are available to help summarize and understand complex spatial distributions, including central tendency, dispersion, and clustering (Cromley and McLafferty, 2002; Rushton, 2003).

Remote Sensing

Remote sensing encompasses the full array of technologies for data collection using aircraft or satellites and includes visible wavelength data as well as a broad range of other types of sensors. It is particularly useful for data describing land use, soil, and hydrological features. Satellite imagery is available over an increasing number of wavelengths and at increasing levels of resolution. Remote sensing, coupled with GIS, has been used widely to describe the environmental conditions associated with disease and to model the occurrence of disease, particularly infectious diseases that are sensitive to environmental conditions such as vectorborne and waterborne diseases.

Data Layers

Data layers are a basic element of GIS. A layer of population data may be superimposed on geological data for determining, for example, whether there is a relationship between bedrock type and population characteristics. Or earthquake vulnerability coefficients may be overlaid on layers showing the distribution of elderly or handicapped people for scenario planning for disaster response. Similarly, maps of land use may be overlaid on digital terrain maps in coastal areas, for example, to aid in efforts to mitigate the salinization problems that have been experienced in Sri Lanka and Malaysia following the flooding of rice paddies by the December 2004 tsunami. Njemanze et al. (1999) used a series of "probability layers" to assess the risk of diarrheal disease from water in rural Nigeria (see Box 7.1). The aggregate risk is a product of geological, hydrological, population, and pollutant characteristics, all of which vary spatially.

BOX 7.1
GIS Data Layers and Diarrheal Disease in Nigeria

Diarrheal diseases are a major cause of mortality and morbidity in developing countries. The spatial distribution of severe diarrhea can be predicted, in part, as a function of the spatial distributions of geological features, population density, and environmental pollution (see Figure 7.1). Population density is important because, other parameters being equal, higher population density tends to increase the rapidity of the spread of disease and also causes an increased number of people to be infected (Halloran, 2001).

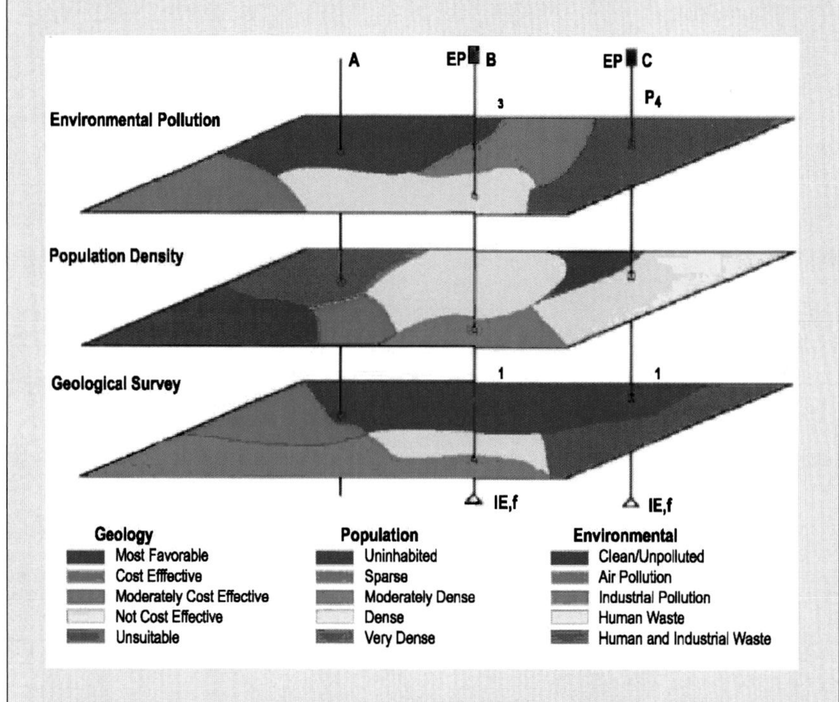

FIGURE 7.1 An example of GIS data layers showing environmental pollution and population density data superimposed on geological features to provide information for understanding the distribution of diarrheal disease in Nigeria.
SOURCE: Njemanze et al. (1999).

Spatial Epidemiology

Any spatially distributed data can be analyzed using spatial statistics, and "spatial epidemiology" has developed as a subfield of epidemiology. Spatial approaches to understanding disease are now feasible because "the availability of geographically indexed health and population data, and advances in computing, geographic information systems, and statistical methodology, have enabled the realistic investigation of spatial variation in disease risk, particularly at the small-area level. Spatial epidemiology is concerned both with *describing* and with *understanding* such variations" (Elliott et al., 2000, p. 3). These authors suggest that there are four types of largely statistical and mathematical studies that fall under the rubric of spatial epidemiology—disease mapping, spatial correlation studies, risk assessment relative to point and line sources, and disease cluster detection. In addition, causal modeling could be added to this list.

TYPES AND AVAILABILITY OF EPIDEMIOLOGICAL DATA

Research at the interface of public health and the earth sciences is only as good as the data used to integrate the two. A wide range of geographical and geological data types, particularly remotely sensed data from earth observation satellites (e.g., Guptill and Moore, 2005), are readily available. Such data are, by definition, spatially distributed, and these data are geocoded to enable spatial modeling, geospatial analysis, and the use of GIS. The same cannot be said of most readily available epidemiological data.

The use of spatial techniques, including GIS and spatial analysis, requires that health data be available with their spatial coordinates. Although health data could be geocoded using either Universal Transverse Mercator spatial coordinates or simply the patient's address, this has implications for the maintenance of privacy of an individual's health status. In addition, a disease with a long latency or highly specific spatial data invites spatial artifacts (e.g., associating a disease with a residential location would be misleading for a work-related illness). In the reasonable absence of such specific information, data could be made available at the census block group or census tract levels. In reality, most health data—if available at all by location—are usually released by zip code (e.g., CHARS data for Washington State, which include detailed diagnostic and procedural information for each patient discharged from a hospital in the state) or by county (e.g., HIV/AIDS data from the Centers for Disease Control and Prevention). The CDC is very concerned with confidentiality, and the concern is that if data are released at a finer scale it may be possible to identify individual patients.

Restricted access to individual health data necessarily makes detailed analyses of spatial patterns of disease more challenging, but it is an almost unavoidable consequence of privacy concerns. This restriction can be addressed in different ways. Seiler et al. (1999) evaluated the opportunity for septic contamination of groundwater by pharmaceuticals by using the specific spatial locations of groundwater wells and linking specific health conditions to individuals through their prescriptions. These authors were able to maintain individuals' privacy by simply refraining from publishing well location data. In other cases, the spatial location (i.e., home address) of an individual may be known by the treating physician or responsible health official, but they would be prohibited from releasing the information (although curiously enough, addresses and even dispatch type for public safety responses involving ambulance or law enforcement are often published in community newspapers). Undoubtedly, making spatial attributes of epidemiological data available for research at appropriate scales and with patient privacy safeguards will continue to pose a challenge. One solution may involve the definition of a new data block of sufficiently small geographic size to be able to associate disease with geological phenomena while providing a sufficiently large error ring around an individual's residence.

Federal Health Datasets

Federal agencies are increasingly using GISs at the interface of the earth sciences and public health. Examples include the Agency for Toxic Substances and Disease Registry (ATSDR), which was an early adopter of GIS (Cromley, 2003), and the Environmental Protection Agency(EPA) with its Toxics Release Inventory[2] (TRI). Although the TRI is not linked to disease data, there is potential to link to cancer registry data, asthma incidence and prevalence data, and other disease data that are spatially distributed. GIS has also been widely used for describing the distribution of natural hazards, for infectious disease modeling and outbreak investigations, for the detection of communicable disease clusters, and—with the recent concern about biowarfare—in new syndromic surveillance systems.[3] Standard datasets collected by the National Center for Health Sta-

[2]See *http://www.epa.gov/tri/*.

[3]For example, see *http://www.syndromic.org/index.php*.

tistics (NCHS) include the National Health Interview Survey[4] (NHIS), the Behavioral Risk Factor Surveillance System,[5] the National Health and Nutrition Examination Surveys[6] (NHANES I, II, etc.), and the National Ambulatory Care Survey.[7] In general, these datasets are available with very poor spatial resolution, although under certain stringent conditions these data may be provided by the NCHS Research Data Center at the county level or, occasionally, the individual level.

Ver Ploeg and Perrin (2004) present a tabulation of available health survey data, listing health outcome data with particular application to social disparities in health but also including most data that are available. For example, the NHANES datasets, representing multiple cross-sectional household surveys, have yielded a great deal of valuable nutritional, cardiovascular, dental, and other data. However, these studies are not available with any geographic specificity and are thus of limited use for understanding any earth science relationship with public health issues. Although NHANES could potentially address the question of whether there is a link between water hardness and cardiovascular disease, this is not possible in the absence of geographically specific data. Another example is the NHIS, which surveys the population comprehensively for major self-reported health conditions. Although the relationship between elevation and hypertension is a potentially interesting question, this dataset cannot be used to address the question because the data are not geographically specific.

A final example is the Adult Blood Lead Epidemiology and Surveillance Program[8] (ABLES), administered by the CDC and the National Institute for Occupational Safety and Health, which measures lead concentrations to estimate risk. Research is currently being carried out to determine the relationship between soil lead levels and lead levels in individuals. However, because the ABLES system does not record geographic data, it cannot be used to address this important research objective.

These surveys and datasets, some of which have large sample sizes and are publicly available, contain extremely valuable data describing health status and diagnoses, health behaviors, diet, risk factors, and other information. However, they are released using the Census Bureau's regional designations (e.g., Northeast, West, South, Midwest), which are

---

[4]See *http://www.cdc.gov/nchs/nhis.htm.*

[5]See *http://www.cdc.gov/brfss/.*

[6]See *http://www.cdc.gov/nchs/nhanes.htm.*

[7]See *http://www.cdc.gov/nchs/about/major/ahcd/namcsdes.htm.*

[8]See *http://www.cdc.gov/niosh/topics/ABLES/ables.html.*

BOX 7.2
Availability of Cancer Data for Spatial Epidemiology

There are a number of problems associated with attempts to establish causality between environmental exposure and cancer data. The National Cancer Atlas includes only mortality data, rather than incidence data, and these data are available only over long time periods and at large units of aggregation. Because environmental exposure to carcinogens occurs at the local level, the National Cancer Atlas is of little use for linking cancer mortality to environmental parameters. The latency period between exposure and detection of the cancer also presents problems, as available data do not permit the reconstruction of life histories and migration histories to trace where exposures may have occurred decades earlier. The "cancer clusters" reported to local and state health departments, the CDC, and the press are often questionable because of the long exposure and latency periods and the absence of life history and migration data.

Another source of cancer data is cancer registries, which contain datasets of individuals in each state together with tissue diagnoses of each histological type of cancer. Incident cases are reported to cancer registries by hospital pathologists based on their tissue diagnoses, so the registries represent the highest degree of diagnostic accuracy that is available. To maintain anonymity, cancer registry data are usually released only at the county level, although it is sometimes possible to obtain data at a more local scale by cancer registry staff or established cancer researchers under strict controls. Gaining access to cancer data usually means sacrificing geographic specificity, and it may also involve including staff members from the cancer registry as coinvestigators. It may then be possible to use geocoded data, at least to the city block, with appropriate safeguards to ensure patient anonymity.

geographically meaningless (e.g., Oklahoma, Indiana, and South Dakota are all included in the "Midwest" despite their lack of similarity in environmental and public health characteristics). Thus, it is impossible to conduct meaningful spatial analysis or mapping of the data from these otherwise very valuable data sources.

Health data in the United States, then, are available from a patchwork of sources and at a variety of nonuniform scales (see Box 7.2). Much health data are available at scales that make it extremely difficult to link to environmental exposures or to conduct spatial analyses. Therefore, researchers are frequently faced with the compelling need to generate primary data at considerable cost.

Scale Issues in Spatially Referenced Health Data

The appropriate spatial scale will vary with the frequency of the disease or health event being analyzed. It is far easier to collect data at a fine scale and aggregate upward than it is to collect data at a large scale and then be forced to infer rates at a smaller scale. Because most health data are available only at a high level of spatial aggregation (e.g., county, zip code, or census tract level) and a great deal of within-unit spatial variation is typically present in data attributes, full spatial analysis is not feasible. For example, the National Cancer Atlas[9] allows queries by state, county, or State Economic Area based on cancer location and occurrence interval. Although there is substantial geographic variation in cancer mortality within states and within counties, this is not reflected in National Cancer Atlas data. Cancer registries in the United States release data only at the zip code level, and because zip codes are arbitrary units with no inherent geographic or geological significance, they are inferior to census tracts or census block groups for demonstrating spatial variation and drawing conclusions with respect to social and economic variations in health disparities (e.g., Krieger et al., 2002). Although staff members at individual cancer registries and some researchers may—under very restrictive conditions—be able to gain access to spatially more specific locations of patient residences, such access is highly variable and depends on study protocols and Institutional Review Board[10] (IRB) restrictions.

Data Access Issues

Why are data so fragmentary and why is it so difficult to obtain data at a fine scale? The fragmentation of data has its roots in government structure, with responsibilities for data collection divided among local health departments, state health departments, and the federal CDC. For conditions—such as cancers—that do not need to be reported to the CDC via local and state health departments, reporting is to local cancer registries. Similarly, trauma cases are reported voluntarily to local trauma registries. Consequently, there is no central repository of health data in the United States and there is considerable variation in the formats and location requirements of the data that are reported.

The reason that the location of incident cases is so difficult to obtain

---

[9]See *http://www3.cancer.gov/atlasplus/*.

[10]IRBs are groups established by individual institutions (universities, private companies, etc.) with the charge to review research to assure the protection of the rights and welfare of the human subjects involved in biomedical research.

stems partly from the fragmentation of data, partly from the fact that many conditions are never reported, and partly from the responsibility of government entities to protect patient confidentiality. There is a fear that revealing specific locations, even to reputable researchers under IRB scrutiny, could compromise the privacy of individual patients. In the case of HIV/AIDS, at least early in the epidemic, the CDC expressed concern that identifying a town as a "hotspot" could result in stigmatization of that town. However, it has never been demonstrated—and is, in fact, implausible—that individuals would be identifiable from data collected at the block scale or the census tract scale. This is particularly the case if data are released only to qualified researchers who have passed appropriate training courses and have no inherent interest in identifying individuals. The Health Insurance Portability and Accountability Act (HIPAA) was designed, in part, to ensure that mandatory standards are established to safeguard the privacy of individually identifiable health information (Hobson, 1997; HHS, 2006)—so far, HIPAA seems to have imposed little constraint on biomedical and epidemiological research. The analytical focus for GIS analysis is on aggregate data patterns rather than on a single data point at a specific location.

The lack of available data and a concern for the environmental sources of disease led to an important report by the Pew Environmental Health Commission (Pew, 2000) that made a strong case for a national environmental health "tracking network" to link environmental sources of disease with resulting health conditions (see Box 7.3). The EPA and the Department of Homeland Security signed a Memorandum of Understanding in 2004 to move in the direction of coordinating data to establish such a system. It is crucial that such a system include geographically referenced health data.

OPPORTUNITIES FOR RESEARCH COLLABORATION

Both infectious and noninfectious diseases vary geographically at scales ranging from very local to global. Some of this variation may be random, and there are inferential tests of spatial randomness. For the variation that is not random, the reasons for that variation include environmental factors. One of the major purposes of GIS, remote sensing, and spatial analysis is not only to describe the variation but also to explain it in terms of environmental variables. This requires that earth and public health scientists collaborate to develop spatially and temporally accurate models for predicting disease distribution that incorporate layers of geological, geographic, and socioeconomic data.

Research to link earth science and public health in the United States is

BOX 7.3
Data Access and Spatial Analysis of Environmental Contamination, Kolding Town, Denmark

Poulstrup and Hansen (2004) used GIS and exposure assessment to investigate the spatial relationship between malignant cancer incidence and exposure to airborne dioxin (Figure 7.2), as a test of the utility of using spatial analysis techniques to assess health effects in a population exposed to environmental contamination. The ability to apply such techniques was dependent on the availability of health and demographic data at the appropriate scale. Health data were derived from the Danish Cancer Registry on an individual basis. The demographic data described each address location (with accuracy to a few meters) and the date of birth, sex, migration (into, out of, and around the area), and date of death for individuals at these addresses.

FIGURE 7.2 GIS output showing spatial relationship between three dioxin sources (red dot), airborne exposure model results (yellow/pink shading), and cancer occurrences (yellow dots) in Kolding Town, Denmark. The green dots are addresses that have been geocoded with Universal Transverse Mercator coordinates with a precision of a few meters and which are associated via Denmark's Central Population Register with each individual's date of birth, sex, migration (into, out of, and around the study area), and date of death.
SOURCE: Poulstrup and Hansen (2004).

severely hampered by the limited availability of geographically referenced, geocoded health data. It is further hampered by the fragmentary nature of many of the available datasets, which are not coordinated, collated, or concatenated. These issues threaten progress in this area of science and may, in the long run, exacerbate disease that results from human-environment interactions. Accordingly, the committee suggests that:

1. **There should be improved coordination between agencies that collect health data, and health data should be merged to the greatest degree possible and made available in formats that are compatible with GIScience analysis.**

2. **Creative solutions to existing restrictions on obtaining geographically specific health data should be investigated, with the goal of defining a geospatially relevant pixel definition that allows predictive and causal analysis while maintaining individual patient privacy. Data made available by federal, state, and county agencies should be geocoded and geographically referenced to this scale.** Legitimate concerns over confidentiality could be further addressed by restricting the release of data to investigators operating under the oversight of Institutional Review Boards.

8

Encouraging Communication and Collaboration

There is broad understanding that promoting communication and collaboration between disparate disciplines is a difficult but not intractable issue facing the full range of natural science, health science, and engineering disciplines. Similarly, there is general acceptance of the considerable potential for synergies from interdisciplinary interactions to result in innovative and exciting research that can lead to new discoveries and greater knowledge. A number of recent National Research Council (NRC) studies have addressed this issue and the following commentary and suggestions are based in part on these analyses (e.g., NRC, 2004d, 2005). This discussion pays particular attention to the earth science and public health communities.

Emphasizing the importance of bridging the "interdisciplinary divide" will not, by itself, promote communication and collaboration—the following discussion focuses on providing suggestions that can realistically be implemented, particularly in the constrained fiscal environment that is likely to apply. It is emphasized that elucidating as yet unrecognized geo-environmental threats to human health may allow avoidance or substantial amelioration of public health problems, thus in the long run saving more money than would have been required for medical treatment.

Encouraging interdisciplinary research requires attention to both the constraints imposed by cultural factors, particularly but certainly not exclusively apparent in traditional discipline-based academic institutions, and to the potential for funding mechanisms to be used by agencies as "carrot and stick" incentives to break down the barriers. The following discussion briefly reviews the existing situation, and then provides suggestions for mechanisms that have the greatest potential for encouraging communication and research collaboration at the interface of earth science and public health.

EXISTING RESEARCH ACTIVITY AND COLLABORATIONS

The key research stakeholders and organizations include individual researchers in academic institutions and private industry as well as a range of federal agencies that either undertake research themselves and/or fund external research. It appears that little research is undertaken by state and local agencies at the intersection of earth science and public health, with these agencies predominantly having a regulatory role with the science base for regulation largely based on federal research activities.

In the course of its discussions and deliberations, the committee received briefings from most of the agencies and organizations involved in research at the interface of earth science and public health. These agencies have far-ranging missions, varying approaches, and differing levels of involvement with research in the earth science/public health arena. The organizations, agencies, and programs described here are not meant to comprise an exhaustive list but rather are intended to identify the major players, both current and potential, in collaborative earth science and public health research.

The **Centers for Disease Control and Prevention–National Center for Environmental Health** (CDC–NCEH) within the Department of Health and Human Services works toward prevention of illness, disability, and death caused by noninfectious and nonoccupational environmental factors through surveillance, applied research, and outreach. In collaboration with the National Aeronautics and Space Administration (NASA), the NCEH is developing a National Environmental Public Health Tracking Network, to integrate environmental hazard and exposure data with data about diseases that are possibly linked to the environment. The **Agency for Toxic Substances and Disease Registry** (ATSDR), an agency within the CDC, jointly addresses environmental public health threats with the NCEH. The ATSDR is an advisory, nonregulatory public health agency that provides health information and public health assessments for toxic substance exposures. Among the tools the ATSDR uses to carry out its mission are toxicological profiles and exposure and disease registries. The current collection of toxicological profiles covers over 250 toxic substances, including arsenic, cadmium, and radon. These profiles contain health effects, exposure pathways, and chemical and physical information. Its collaborative efforts include mineralogical characterization of fibrous amphiboles with the U.S. Geological Survey (USGS) and Environmental Protection Agency at the vermiculite (asbestos) mine in Libby, Montana.

The **Department of Defense–Armed Forces Institute of Pathology** (DOD–AFIP) specializes in pathology consultation, education, and research. The Department of Environmental and Toxicologic Pathology within the AFIP focuses on techniques for tissue analysis and maintains the INTOX Data Center, which includes the Medical Geology Database and Chronic Arseniasis Database. The Medical Geology Database contains information about sources of harmful materials in the environment, including exposure pathways and prediction of the movement of disease-causing agents. The AFIP has collaborated with USGS in the past, examining environmental problems, including arsenic exposure, caused by a 1996 tailings spill from an open-pit copper mine on Marinduque Island in the Philippines (Plumlee et al., 2000).

The **Environmental Protection Agency** (EPA) undertakes a variety of federal research, monitoring, standards setting, and enforcement activities to ensure environmental protection. Broadly, EPA's mission is to protect human health and safeguard the natural environment. One collaborative effort currently under way is the Superfund Basic Research Program (SBRP), which is funded by the National Institute of Environmental Health Sciences and coordinated with the EPA. The SBRP funds university-based multidisciplinary research on public health and remediation technologies at hazardous waste sites. The research supported by the program encompasses many fields, including chemistry, ecology, epidemiology, toxicology, molecular biology, hydrogeology, engineering, and soil science. EPA has also provided a list of emerging infectious agents that are of concern for waterborne transmission—the so-called Contaminant Candidate List (EPA, 2003).

The **National Aeronautics and Space Administration** (NASA) performs a broad range of space-based research on the earth that includes a specific public health component within a broader environmental monitoring program. NASA collects data and funds external research aimed at enhancing decision support tools using observations and modeling of weather, climate and other environmental factors that influence disease vectors and air quality. NASA has partnered with the CDC to enhance the National Environmental Public Health Tracking Network, which uses estimates of ground-level ozone, particulate matter, and/or other atmospheric pollutants to provide warnings of increased risk of respiratory diseases such as asthma and emphysema. Other public health collaborations involve DoD, EPA, NIH, NOAA, and USGS.

The **National Institutes of Health** (NIH) contains several organizations under its umbrella that conduct and support research at the earth science/public health interface. At the institute level, the **National Insti-**

tute of Environmental Health Sciences (NIEHS) focuses on basic science, disease-oriented research, global environmental health, and multidisciplinary training for researchers. It both supports and conducts research and training in addition to health information outreach and communication programs. Its internal research programs include the Environmental Diseases and Medicine Program and the Environmental Toxicology Program. The first program focuses on diseases and physiological dysfunctions that have known or suspected environmental components in their etiologies, with an emphasis on cancer, reproductive and developmental dysfunction, and pulmonary diseases; it also plans and conducts epidemiological studies. The latter program supports the National Toxicology Program by providing evaluations of toxic substance of public health concern (e.g., NTP, 2005) and strengthening risk assessment approaches and data. The NIEHS has collaborated with the USGS on the Environmental Mercury Mapping, Modeling, and Analysis program, and with the EPA on the SBRP.

The **National Cancer Institute** (NCI) is the federal government's principal agency for cancer research and training. The NCI both supports and conducts research, training, and health information dissemination relevant to the cause, diagnosis, prevention, and treatment of cancer, as well as rehabilitation. Under the premise that most cases of cancer are linked to environmental causes and, in principle, can be prevented, the NCI's Division of Cancer Epidemiology and Genetics is working with NIEHS to address the contribution of various agents, including exposure to those in air and water, to the nation's overall cancer burden.

At the center level, the NIH also contains the **John E. Fogarty International Center for Advanced Study in the Health Sciences,** the international component of the NIH that addresses global health challenges. The center is the primary NIH partner in the joint NIH-NSF Ecology of Infectious Diseases initiative, an excellent example of an existing earth science and public health collaborative research program (see Box 8.1).

The **National Science Foundation** (NSF) sponsors basic research encompassing the full range of science and engineering disciplines. Both the Geosciences Directorate and the Biological Sciences Directorate support research relevant to earth science and public health, with the goal of describing both the positive and negative connections between the two areas over the full range of scales. Although NSF's mission specifically excludes the medical sciences, the foundation does support interdisciplinary research with NIH, including the Ecology of Infectious Diseases initiative and the Oceans and Human Health Research Centers (see Box 8.2).

The **United States Geological Survey** of the Department of the Interior supports applied earth and natural science research by its own researchers and funds external research activities. The USGS currently conducts a number of health-related activities to improve understanding of the links between human health and geological processes, including research on the distribution and health effects of asbestos, cadmium, chromium, lead, mercury, radon, selenium, and uranium; water quality monitoring; hazard forecasting; and bacterial and viral transport in groundwater. The USGS has the mandate to carry out national soil mapping, and a detailed map of the nation's soil resources could form the geochemical framework for a significant component of the priority research at the interface of earth science and public health noted in this report. The USGS also supports the Human Health Database,[1] which provides information about the national distribution of arsenic, radon, and mercury as well as land cover datasets and mineral resources spatial data. In collaboration with NIH–NIEHS, the USGS created the Environmental Mercury Mapping, Modeling, and Analysis website to support environmental and health care researchers as well as land and resource managers. Although this program does not fund research at either the agency or academic level, it provides consolidated information in the form of maps and data for research support.

MODELS FOR ENCOURAGING COLLABORATIVE RESEARCH

The value of interdisciplinary research has been convincingly documented in another NRC report (NRC, 2004d), which provides a comprehensive description of the barriers to collaborative research and suggests strategies for overcoming these barriers. For the specific case of earth science and public health, a variety of activities could be supported to further the implementation of interdisciplinary research agendas including:

- Funding of new interdisciplinary collaborative centers. One approach with the potential to encourage true collaboration is the "Center-Based Approach" epitomized by the NIEHS Superfund Basic Research Program (SBRP). In this collaborative model, diverse groups of scientists and engineers are mandated to collaborate on different aspects of specific issues. Major funding is provided to approximately 20 universities nationally, with each university focusing on a specific set of issues involving the cleanup or remediation of superfund sites. To be successful, applicants must have both toxicology components and environmental science components. The superfund program has been extremely successful and illustrates that major funding is an effective mechanism to promote true collaboration and innovative approaches to complex issues.

[1]See *http://health.usgs.gov/health_database.html.*

BOX 8.1
Ecology of Infectious Diseases—A Multiagency Initiative

The Ecology of Infectious Diseases (EID) is a joint National Institutes of Health (NIH)–National Science Foundation (NSF) initiative to develop predictive models to describe the dynamics of infectious diseases by supporting research that falls outside the current scope of each agency's mainstream research programs (NIH–NSF, 2005). The EID supports efforts to understand the underlying ecological and biological mechanisms that govern relationships between human-induced environmental changes and the emergence and transmission of infectious diseases. The highly interdisciplinary research projects funded by this program examine how large-scale environmental events—such as habitat destruction, biological invasion, and pollution—alter the risks of emergence of viral, parasitic, and bacterial diseases in humans and other animals. The initiative is administered by the Fogarty International Center (FIC), the National Institute of Allergy and Infectious Diseases, the National Institute of Environmental Health Sciences, and the NSF. Examples of research awards made by the EID program since 1999 included "Microbial Community Ecology of Tick-borne Human Pathogens," "Impact of Land-Cover Change on Hantavirus Ecology," and "Environmental Determinants of Malaria in Belize."

Between 1999 and 2005, 42 research awards were made under the EID initiative, with total funding of approximately $60.7 million. In fiscal year 2006, the program's sixth year of funding, $8 million was available for new awards, made up of $6.5 million provided by NSF and approximately $1.5 million from NIH. Although funding levels have increased since the program's inception in 1999, the proportion of funding from NIH and NSF has changed (see Figure 8.1); funding amounts from the two agencies were

- Resource and infrastructure enhancements (facilities, equipment, databases and other central resource libraries and services, research training through institutional training programs or individual fellowships, etc.), when specifically established to cross disciplinary boundaries.
- Funding of individual research projects (through awards, grants, cooperative agreements, contracts), when such projects are designed to cross disciplinary boundaries (e.g., the Ecology of Infectious Diseases initiative; see Box 8.1).

Funding Collaborative Research

In an environment of abundant suggestions and ideas for collaborative research but scarce resources, funding mechanisms hold the key to

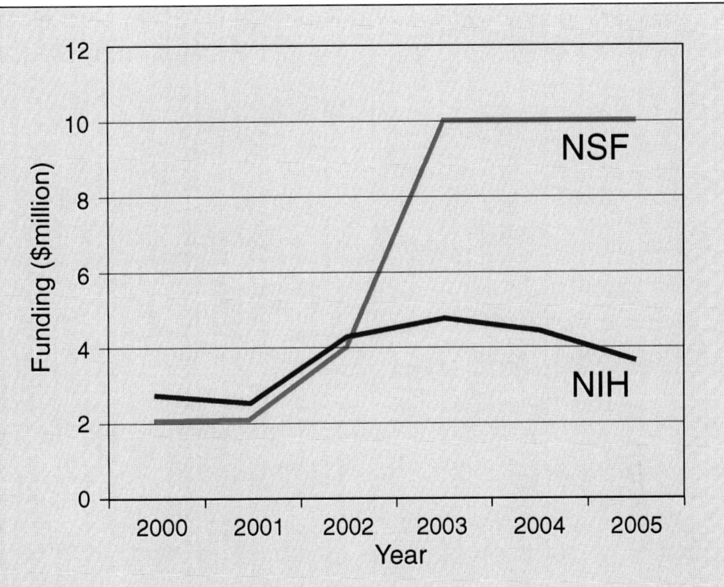

FIGURE 8.1 Graph showing funding provided by NSF and NIH for the Ecology of Infectious Diseases initiative between 2000 and 2005.

approximately equal in 2000, but by 2005 the NSF was funding the program at more than twice the level of NIH. Despite funding challenges, some of which stem from the interdisciplinary nature of the research (NIH–NSF, 2005), the EID program provides a model for collaborative activity between the health and earth sciences.

successful establishment of research collaborations. Many funding agencies issue Requests for Proposals (RFPs) that solicit research within specific areas or issues. Frequently, this results in individual applications by one or several faculty members at a single institution. Although multidisciplinary approaches are often specifically encouraged in such RFPs, true collaboration across disciplines is rare. Research collaborations can be based on co-funding, complementary funding, or single-agency funding. Co-funding involves shared funding of a single project or activity by two or more organizations (e.g., the Microbial Observatories initiative funded by NSF and the U.S. Department of Agriculture); complementary funding involves separate funding by two or more organizations of individual projects or activities that relate to each other (e.g., the EID initiative funded by NSF and NIH–FIC); and single-agency funding is

BOX 8.2
Oceans and Human Health Research Centers—
A Multiagency Initiative

NSF and NIEHS provided funding in 2004 for the establishment of four joint Centers for Oceans and Human Health to facilitate collaborative biomedical and oceanographic research, with the expectation that combined agency funding will be $5 million annually for five years. These centers, and the special programs on which they will focus, are located at the University of Washington (toxic algae and human health effects of contaminated shellfish), the University of Hawaii (microbial pathogens in tropical coastal waters and potential pharmaceutical uses for extracts from tropical microorganisms), Woods Hole Oceanographic Institution (variations in populations of the toxic plankton *Alexandrium* and its effect on shellfish toxicity), and the University of Miami (harmful algal blooms in subtropical ecosystems and microbial components of coastal water and their effects on human health).

where multiple collaborative research centers encompassing a variety of disciplines are funded by one agency (e.g., SBRP funded by NIEHS).

Role of Academia in Encouraging Collaborative Research

Traditionally, academic research has been based on the *individual* efforts of scholars at private and public universities. Throughout the learning process of graduate school, postdoctoral studies, and the ultimate prize of obtaining an assistant professorship, scholars are trained to be unique. As assistant professors struggle to obtain promotion and tenure during their first several years at a university, they seek a niche that distinguishes them and their programs from other faculty. The culture of academia rewards individual scholarship, and as a consequence individual scientists tend to be highly competitive. In addition, even though students take a diverse array of classes during graduate school, they tend to specialize in well-defined and focused disciplines. Because of these characteristics, collaborative research has been neither particularly encouraged nor rewarded. Institutional barriers between different academic departments, and the tendency for academic departments to recruit new staff with traditional disciplinary expertise, work to dissuade faculty from collaborative interdisciplinary research. For faculty to collaborate effectively in multidisciplinary research, specific criteria must be met. First, the faculty members must be personally compatible. Second, the areas of expertise must be complementary, permitting a symbiotic relationship. And third, funding must be obtained to allow for the collaborative research to be initiated. The first two factors are controlled by faculty themselves, but the third factor is controlled by funding agencies. They, too, are competitive, and institutional and often personality barriers to collaborating with other funding agencies exist.

The distinction between fundamental basic research and applied research has been well documented but perhaps overemphasized. Traditionally, academics have perceived that basic research is more prestigious than applied research, particularly if the applied research has been funded by the private sector. However, in reality, most basic research is ultimately utilized for the solution of applied problems. Therefore, faculty themselves can engage in collaborative research by pursuing the integration of basic and applied research. In many instances, collaboration across scientific disciplines, and in particular between basic scientists and engineers, has proved to be an effective mechanism to provide a holistic solution to particular problems. To encourage this requires both a culture change and the education of a "new breed" of graduate students, trained to not only specialize in a given field but also acquire a broader based background relevant to their field of interest.

For example, microbial activity in soils is heavily influenced by the chemistry and physical matrix of the environment—the fate and transport of viruses through soil is dependent not only on the size and biological properties of the organism itself but also on the physical and chemical properties of the soil. To study the fate and transport of viruses through soil, collaboration between soil physicists and microbiologists is essential. However, there is little incentive to undertake such collaboration. Although it would be appropriate for scientific curiosity and innovation to provide this stimulus, in reality the acquisition of funding for such collaborative research is the real driving force. Therefore, although a broad range of recommendations and suggestions applying to academic institutions can be listed—for example, the range of recommendations included in NRC (2004d)—it is clear that funding agencies must play a critically important role in implementing collaborative research.

Role of Federal Agencies in Encouraging Collaborative Research

The purviews of the many federal agencies with responsibilities related to earth science and public health are reasonably well defined. While a number of these agencies can be considered as having peripheral involvement at the intersection of these fields, it is clear that there are several agencies—specifically the CDC, DoD, EPA, NASA, NCI, NIEHS, NSF, and USGS—that have most of the responsibility for research at this disciplinary intersection, and it is these agencies that are the primary focus of this committee's analysis and recommendations. Ultimately, it is the program managers within these agencies—who support research activities within the organization or fund external research at academic institutions, government laboratories, and state agencies—that must be convinced of the value of promoting collaborative research at the intersection of earth science and public health and be provided with recommendations concerning priority research areas and the optimum means for facilitating collaborative research.

All of the cultural barriers that exist in academic institutions have their parallels in funding agencies. In practice, the peer review systems used by most agencies mean that each additional disciplinary area involved in a particular proposal in effect equates to an additional hurdle that must be surpassed, so that more restricted proposals are widely understood to have the highest chance of success. The metrics for assessing the success of the programs overseen by individual managers are almost always based on individual program funding levels and provide little incentive for valuing the "discounted research" that can result from partial funding of cross-program activities. On the broadest scale, the fact that the appropriation levels for these agencies are determined by a range of congressional appropriations committees—which themselves have no intrinsic incentive to promote cross-committee activity—means that a top-down insistence on interdisciplinary research is unlikely. While there is broad, and even in some agencies a pervasive recognition of the merits of interdisciplinary research, the existence of barriers and the lack of incentives mean that the focus here must be on suggesting practical measures, with incentives, that can be taken by individual program managers and their superiors.

To date, support for collaborative research *between* the relevant agencies has been limited, although increasing (but still minor) support for collaborative research *within* specific funding agencies suggests that models for such research are being developed, and experiences gained, that should be applicable to cross-agency collaboration. Existing collaborative research models include NSF-based programs such as the Integrative Graduate Education and Research Traineeship (IGERT) and Biocomplexity in the Environment initiatives, and NIH-based collaborative programs such as the NIH Road Map for Interdisciplinary Research.

The classic risk assessment paradigm (see Box 8.3) provides an illustration of the need for cross-agency research support. Two of the key factors in risk characterization are exposure assessment and dose response. Historically, funding for exposure assessment involving environmental science research came from NSF or EPA, whereas funding for dose-response studies came from NIH. It seems obvious that such risk character-

BOX 8.3
The Risk Assessment Paradigm

Risk assessment consists of four elements:

- **Hazard identification**—defining the hazard and nature of the harm; for example, identifying a chemical contaminant, say lead or carbon tetrachloride, and documenting its toxic effects on humans.
- **Exposure assessment**—determining the concentration of a contaminating agent in the environment and estimating its rate of intake in target organisms; for example, finding the concentration of aflatoxin in peanut butter and determining the dose an "average" person would receive.
- **Dose-response assessment**—quantifying the adverse effects arising from exposure to a hazardous agent based on the degree of exposure. This assessment is usually expressed mathematically as a plot showing the response in living organisms to increasing doses of the agent.
- **Risk characterization**—estimating the potential impact of a hazard based on the severity of its effects and the amount of exposure.

ization can most effectively be accomplished by linked research using some model that provides pooled resources from all these agencies.

MULTIAGENCY SUPPORT FOR COLLABORATIVE RESEARCH

Despite the absence of existing institutional mechanisms to support research activity at the interface of earth science and public health, a base level of collaborative research exists primarily as a consequence of individual scientists establishing research partnerships and individual program managers in a variety of agencies identifying the strong merits of such research and providing support. It is also clear that there is a considerable amount of existing research activity in both the earth science and public health fields that is being carried out without knowledge or communication of potentially complementary research "across the divide." the committee suggests that, for there to be substantial and systemic advances in interdisciplinary interaction, a formal multiagency collaboration support system needs to replace the existing ad hoc nature of collaborations. Within this context, and despite wariness about proposing yet another bureaucratic structure, the committee believes that a useful contribution would be to suggest a multitiered hierarchical management and coordination mechanism by which the relevant funding agencies could interact to promote communication and collaboration.

The premise upon which this suggested structure is based is the belief that there are senior managers in most of the relevant agencies who acknowledge the value of collaborative research activity. However, within the context of the existing distributed appropriations mechanisms, it is highly unlikely that direct single-line funding for collaborative activities will be provided. Accordingly, the committee proposes a research support structure (see Figure 8.2) that consists of a policy-oriented direction and coordination council and an implementation-oriented steering committee, with the following roles:

Coordination Council. This should consist of senior managers from each of the primary agencies identified above (CDC, EPA, NASA, NCI, NIEHS, NSF, and USGS). This council would focus on the policy aspects of collaborative activity and would include identifying broad research areas, securing commitments from each agency, and identifying the research gaps that need to be filled. The Committee on Environment and Natural Resources of the National Science and Technology Council, administered within the Office of Science and Technology Policy, might provide an appropriate forum or mechanism for the cross-agency collaboration that is required.

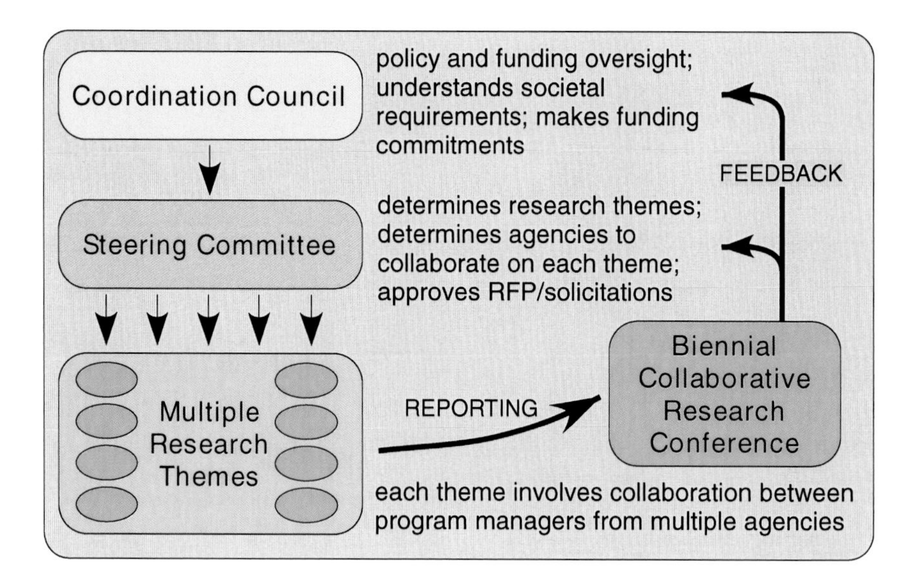

FIGURE 8.2 Proposal for a research support mechanism for interagency collaboration.

Steering Committee. This committee would consist of program managers from the agencies represented on the coordination council and other agencies with related interests. For example, a relevant theme in the earth science/public health arena might focus on trichloroethylene (TCE) contamination of groundwater. RFPs in this area might solicit research designed to (1) evaluate the health effects of TCE on human health; (2) determine human exposure levels based on bioavailable TCE in groundwater at specific sites and consumption of water; and (3) technology to remove TCE from groundwater. Clearly, in this example, program managers from EPA, NIEHS, NIH, NSF, and USGS should be involved. This steering committee would approve specific research themes (not projects) and would also develop specific RFPs. Individual program managers from different agencies would need to interact to approve the funding of multiple projects that together make up a specific research center (which might be located at one institution or distributed among several). In addition, the success of a particular research center would require evaluation by a multiagency group of managers.

Biennial Collaborative Research Conference. An earth science and public health collaborative research conference should be held every two years, attended by the coordination council, steering committee, and representatives of the research themes. The purpose of this meeting would be to evaluate the success of collaborations over the previous two years and also to propose high-priority research collaborations that would be the basis for RFPs for the next four years.

With the understanding that a certain level of commitment is required before a research collaboration can be considered a true collaborative activity, the committee has arbitrarily established a 20% contribution as a minimum (e.g., we do not consider a 95:5 funding split as constituting a real collaboration). A funding system is envisioned wherein more than one agency provides support for multiple individual research projects, with RFPs/solicitations and award reporting coordinated by the steering committee but administered according to each individual agency's rules.

9

Collaborative Research Priorities

Human health depends on both intrinsic characteristics and external biotic or abiotic environmental exposures that may occur via the pathways of air, water, food, or contact. Although many adverse exposures stem from anthropogenic disturbances of the natural environment, others come directly from natural sources near or at the earth's surface. Earth sciences—including geology, geophysics, geochemistry, geomorphology, soil science, mineralogy, hydrology, mapping, remote sensing, and other subdisciplines—are concerned with natural earth materials and processes, and knowledge of these materials and processes is an essential component for understanding such adverse exposures.

In the public health paradigm, early disease prevention, rather than control and treatment, affords good health to the greatest number of individuals. With increased emphasis on prevention, understanding disease-causing environmental exposures is essential for cost-effective improvements to human health—the recognition and resultant avoidance of bioassimilated earth materials potentially injurious to human health is likely to save considerably more money than would be required to treat the adverse public health effects caused by the ingestion, respiration, and/or absorption of such materials (e.g., savings in health and mortality costs as a consequence of improved air quality resulting from implementation of the Clean Air Act; EPA, 1997, 1999; NRC, 2002b). Despite such obvious and important links between the fields of earth science and public health, public health activities and research have paid relatively little attention to natural earth materials and processes. Improved integration of these disciplinary fields and communities offers significant potential for an improved understanding of exposure mechanisms and health risks, particularly in an era of extensive utilization of the earth's natural resources.

This report highlights examples of successful research collaboration and emphasizes the limited number of interagency initiatives at the interface of these disciplines. For the most part, government agencies have failed to adequately promote the necessary integration of the earth and health sciences. **The committee is convinced that substantial improvements in public health can be achieved as a result of increased collaboration between the earth science and public health communities.**

RESEARCH THEMES AND PRIORITIES

The research priorities presented here cut across the human exposure pathways described in Chapters 3 through 7 and are illustrated by the particular priority research activities proposed in those chapters. In compiling these recommendations, the committee required that the research proposed must involve collaboration between researchers from both the earth science and public health communities and did not consider the abundant examples of valuable research that could be undertaken primarily within one or other of the disciplines. These recommendations are not listed in any rank order. It is important to note that the multidisciplinary research teams needed to effectively address these priorities will in many cases require the involvement of other specialist disciplines beyond the narrowly defined earth science and public health areas (e.g., atmospheric scientists, environmental engineers).

Earth Material Exposure Assessments: Understanding Fate and Transport

Assessment of human exposure to hazards in the environment is often the weakest link in most human health risk assessments. The physical, chemical, and biological processes that create, modify, or alter the transport and bioavailability of natural or anthropogenically generated hazardous earth materials remain difficult to quantify. A vastly improved understanding of the spatial and geochemical attributes of potentially deleterious earth materials is a critical requirement for effective and efficient mitigation of the risk posed by such materials. **An improved understanding of the source, fate, transport, and bioavailability of potentially hazardous earth materials is a critically important research priority.** Collaborative research should include:

- Addressing the range of issues associated with airborne *mixtures* of pathogens and chemical irritants. The adverse effects arising from the inhalation of complex mixtures of pathogens and chemical and biochemical species in airborne pollution require detailed geological investigations of earth sources and the identification of atmospheric pathways to sites of bioaccessibility and potential ingestion by humans. The anticipation and prevention of health effects caused by earth-sourced air pollution prior to the onset of illness require quantitative knowledge of the geospatial context of earth materials and related disease vectors.

- Determining the influence of biogeochemical cycling of trace elements in water and soils as it relates to low-dose chronic exposure via toxic elements in foods and ultimately its influence on human health. In general, little is known of the elemental interactions and the influence of mixtures of elements on bioavailability in soils; in residential, industrial, and irrigated water supplies; and within the human body.
- Determining the distribution, survival, and transfer of plant and human pathogens through soil with respect to the geological framework. Collaboration would involve earth scientists to characterize the biogeochemical habitat, such as the exchangeable cations, mobile metal species, and/or reactive geochemical surfaces, including sources of nutrients, or the presence of antagonistic and/or synergistic metal species. Microbiologists would characterize the microbial community that surrounds the pathogen and examine its viability in different biogeochemical habitats, and public health specialists would examine the incidence of human and plant disease from soil pathogens as a function of the biogeochemical framework and the role of soils in long-term survival of pathogens and as reservoirs of pathogens.
- Improving our understanding of the relationship between disease and both metal speciation and metal-metal interaction. In this research, earth scientists would characterize metal abundance and metal speciation in water and soils and the mobility and availability of these metals to the biosphere; microbiologists would characterize the microbial populations and mechanisms that are responsible for metal species transitions in water and soil environments; and public health specialists would use spatial information on the distribution of metal speciation to examine the incidence and transfer of specific disease.
- Identifying and quantifying the health risks posed by "emerging" contaminants, including newly discovered pathogens and pharmaceutical chemicals, which are transported by earth processes. The health effects of many naturally occurring substances at low concentrations and the health effects associated with interactions of multiple naturally occurring substances are poorly understood. A particular focus should be the "emerging" contaminants, such as hormones, pharmaceuticals, personal care products, and newly identified microbial pathogens, for which sources and transport processes are poorly understood. The synergistic

and antagonistic interactions of contaminants with naturally occurring substances in water, and similar interactions among multiple naturally occurring substances or multiple contaminants, also pose priority research questions.

Improved Risk-Based Hazard Mitigation

Natural earth processes—including earthquakes, landslides, tsunamis, and volcanoes—continue to cause numerous deaths and immense suffering worldwide. As climates change, the nature and distribution of such natural disasters will undoubtedly also change. **Improved risk-based hazard mitigation, based on improved understanding of the public health effects of natural hazards under existing and future climatic regimes, is an important research priority.** Such collaborative research should include:

- Determining processes and techniques to integrate the wealth of information provided by the diverse earth science, engineering, emergency response, and public health disciplines so that more sophisticated scenarios can be developed to ultimately form the basis for improved natural hazard mitigation strategies. Any assessment of population vulnerability is dependent on the merging of earth science information describing the spatial distribution of hazards with public health information describing population characteristics and medical response capabilities. Effective scenarios to form the basis of improved response strategies must be scientifically valid and believable for broad acceptance by those charged with disaster response planning. The scientific validation will require collaborative involvement of a broad range of experts from the earth science, public health, emergency management, and engineering communities.

Assessment of Health Risks Resulting from Human Modification of Terrestrial Systems

Human disturbances of natural terrestrial systems—for example, by activities as diverse as underground resource extraction, waste disposal, or land cover and habitat change—are creating new types of health risks. **Research to understand and document the health risks arising from disturbance of terrestrial systems is key to alleviating existing health threats and preventing new exposures.** Such collaborative research should include:

- Analysis of the effect of geomorphic and hydrological land surface alteration on disease ecology, including emergence/resurgence and transmission of disease. The effects of land surface modifications on the bioaccessibility of the vectors responsible for disease outbreaks, deleterious chronic disease levels, and human senescence need to be measured individually and collectively in order to better safeguard public health. This will require a seamless integration of geological, hydrological, and epidemiological research efforts.

- Determining the health effects associated with water quality changes induced by novel technologies and other strategies currently being implemented, or planned, for extending groundwater and surface water supplies to meet increasing demands for water by a growing world population. Water stored in a brackish aquifer during aquifer storage and recovery will experience an increase in total dissolved solids due to mixing with ambient brackish water and dissolution of minerals from the aquifer matrix, and this may introduce contaminants such as microbial pathogens, organic contaminants such as pesticides or solvents, and inorganic contaminants such as nitrates or metals. Of particular interest with respect to surface water are changes in water quality induced by urban and agricultural runoff, discharge of waste effluents from municipal or industrial sources (including the extractive mineral and energy industries), construction and operation of dams and reservoirs, drainage of wetlands, and channel modifications for purposes of flood control, navigation, or environmental improvement.

IMPLEMENTATION STRATEGIES

Understanding the importance of bridging the "interdisciplinary divide" will not, by itself, promote communication and collaboration. The following recommendations to facilitate research across the disciplinary boundaries are those the committee believes can realistically be implemented, particularly in the constrained fiscal environment that is likely to apply.

Interdisciplinary Spatial Analysis

Spatial attributes describe the distribution of natural earth materials and processes and the distribution of infectious and noninfectious diseases. The epidemiological analysis techniques commonly used by public health scientists, when linked with geospatial analysis techniques developed by earth scientists, provide a powerful tool for understanding the public health effects of earth science materials. The immense potential offered by modern spatial analysis tools provides a strong impetus to break down existing institutional barriers to making public health data with adequate geographic detail more widely available for research. **The application of modern complex spatial analytical techniques has the potential to provide a rigorous base for integrated earth science and public health research by facilitating the analysis of spatial relationships between public health effects and natural earth materials and processes.**

The committee, therefore, suggests pursuing research that (1) leads to spatially and temporally accurate models for predicting disease distribution based on integrated layers of geological, geographic, and socioeconomic data; (2) develops better technologies for high-resolution data generation and display; and (3) establishes user-friendly Geographic Information Systems (GISs) for the earth science and public health communities and includes GIS and spatial analysis in the training of public health professionals.

Before it will be possible to take advantage of the considerable power of modern spatial analysis techniques, a number of issues associated with data access will need to be addressed. Improved coordination between agencies that collect health data will be required, and health data from the different entities will need to be merged and made available in formats that are compatible with GIScience analysis. Existing restrictions on obtaining geographically specific health data, while important for maintaining privacy, severely inhibit effective predictive and causal analysis. To address this, it will be necessary for all data collected by federal, state, and county agencies to be geocoded and geographically referenced to the finest scale possible, and artificial barriers to spatial analysis resulting from privacy concerns need to be modified to ensure that the enormous power of modern spatial analysis techniques can be applied to public health issues without compromising privacy. The potential of the research community to perform sophisticated multilayered spatial analysis for both predictive and causal modeling requires that epidemiological data be presented in a useful (i.e., detailed geographic) format so that public health problems can be correlated with earth science parameters.

Interagency Support for Interdisciplinary Earth and Public Health Sciences Research

The value of interdisciplinary research has been convincingly and repeatedly described (e.g., NRC, 2004d). However, while important gains have been made *within* individual funding agencies toward interdisciplinary research, a dearth of collaboration and funding *between* agencies has restricted significant scientific discovery at the interface of public health and earth science. The importance of the links between these disciplines and the considerable potential for collaborative earth science and public health research to produce major societal benefits has led the committee to conclude that multiagency funding for collaborative research will achieve greater disease prevention than will traditional single-agency funding. The committee recommends that—in order to achieve the innovative research at the intersection of public health and earth science required for societal benefit and protection—the intellectual and fiscal resources presently existing within a number of agencies should be reallocated to a shared interagency pool to support an interdisciplinary public health and earth science initiative. The ultimate goal is to enhance interdisciplinary collaboration, both across federal agencies and across academic and other research institutions, so that society can benefit from the nation's considerable expertise in earth science and public health.

The interface between the earth sciences and public health is pervasive and enormously complex. Collaborative research at this interface is in its infancy, with great potential to ameliorate the adverse health effects and enhance the beneficial health effects from earth materials and earth processes. The earth science and public health research communities share a responsibility and an obligation to work together to realize the considerable potential for both short-term and long-term positive health impacts.

References

AAAI (American Academy of Allergy Asthma & Immunology), 2000. *The Allergy Report*. Milwaukee, Wisc., AAAI.

Abedin, J.M., J. Feldman, and A.A. Meharg, 2002. Uptake kinetics of arsenic species in rice plants. *Plant Physiology,* 128: 1120–1128.

Abrahams, P.W., 2003. Human Geophagy: A Review of its Distribution, Causes, and Implications. Pp. 31–35 *in* H.C.W. Skinner and A.R. Berger (eds.), *Geology and Health: Closing the Gap*. New York, Oxford University Press, 179 pp.

Abrahams, P.W., 2005. Geophagy and the Involuntary Ingeston of Soil. Pp. 435–458 *in* O. Selinus, B.J. Alloway, J.A. Centeno, R.B. Finkelman, R. Fuge, U. Lindh, and P. Smedley (eds.), *Essentials of Medical Geology*. London, Elsevier Academic Press, 812 pp.

ADA (American Dietetic Association), 2005. Position of the American Dietetic Association: The impact of fluoride on health. *Journal of the American Dietetic Association,* 105: 1620–1628.

Adriano, D.C., 2001. *Trace Elements in Terrestrial Environments: Biogeochemistry, Bioavailability, and Risk of Metals, 2nd edition*. New York, Springer, 867 pp.

Allen, H.E. (ed.), 2002. *Bioavailability of Metals in Terrestrial Ecosystems: Importance of Partitioning for Bioavailability to Invertebrates, Microbes and Plants*. Pensacola, Fla., SETAC Press.

Alloway, B.J., 2005. Bioavailability of Elements in Soil. Pp. 347–372 *in* O. Selinus, B.J. Alloway, J.A. Centeno, R.B. Finkelman, R. Fuge, U. Lindh, and P. Smedley (eds.), *Essentials of Medical Geology*. London, Elsevier Academic Press, 812 pp.

Anderson, M.B., 1991. Which costs more: Prevention or recovery? Pp. 17–27 *in* A. Kreimer and M. Mnasinghe (eds.), *Managing Natural Disasters and the Environment*. Washington, D.C., World Bank.

Andersen, O., and J.B. Nielsen, 1994. Effects of simultaneous low-level dietary supplementation with inorganic and organic selenium on whole-body, blood and organ levels of toxic metals in mice. *Environmental Health Perspectives,* 102 (Suppl. 3): 321–324.

Appleton, J.D., 2005. Radon in Air and Water. Pp. 227-262 *in* O. Selinus, B.J. Alloway, J.A. Centeno, R.B. Finkelman, R. Fuge, U. Lindh, and P. Smedley (eds.), *Essentials of Medical Geology*. London, Elsevier Academic Press, 812 pp.

Arnold, R., D. Quanrud, C. Gerba, and I.L. Pepper, 2006. Pharmaceuticals and Endocrine Disruptors. Pp. 506–516 *in* I.L. Pepper, C.P. Gerba, and M.L. Brusseau (eds.), *Environmental and Pollution Science, 2nd Edition*. San Diego, Calif., Elsevier Science/Academic Press, 532 pp.

Artiola, J.F., I.L. Pepper, and M.L. Brusseau (eds.), 2005. *Environmental Monitoring and Characterization*. San Diego, Calif., Elsevier Academic Press.

Ashbolt, N.J., 2004. Microbial contamination of drinking water and disease outcomes in developing regions. *Toxicology*, 198: 229–238.

Ayotte, J.D., M.G. Nielsen, G.R. Robinson, Jr., and R.B. Moore, 1999. Relation of Arsenic, Iron, and Manganese in Ground Water to Aquifer Type, Bedrock Lithogeochemistry, and Land Use in the New England Coastal Basins. *Water Resources Investigations Report*, 99-4162.

Ayotte, J.D., B.T. Nolan, J.R. Nuckols, K.P. Cantor, G.R. Robinson, D. Baris, L. Hayes, M. Karagas, D.T. Silverman, and J. Lubin, 2006. Modeling the probability of arsenic in groundwater in New England as a tool for exposure assessment. *Environmental Science and Technology*, 40: 3578–3585.

Baris, Y.I., M. Artivinli, and A.A. Sahin, 1979. Environmental mesothelioma in Turkey. *Annals of the New York Academy of Sciences*, 330: 423–432.

Baxter, P.J., R. Ing, H. Falk, and B. Plikaytis, 1983. Mount St. Helens eruptions: The acute respiratory effects of volcanic ash in a North American community. *Archives of Environmental Health*, 38: 138–143.

Bejat, L., E. Perfect, V.L. Quisenberry, M.S. Coyne, and G.R. Haszler, 2000. Solute transport as related to soil structure in unsaturated intact soil blocks. *Soil Science Society of America Journal*, 64: 818–826.

Benin, L., 1985. *Medical Consequences of Natural Disasters*. Berlin, Springer-Verlag, 160 pp.

Bennett, P.C., F.K. Hiebert, and J.R. Rogers, 2000. Microbial control of mineral-groundwater equilibria-macroscale to microscale. *Hydrogeology Journal*, 8: 47–62.

Bennett, P.C., J.R. Rogers, and W.J. Choi, 2001. Silicates, silicate weathering, and microbial ecology. *Geomicrobiology Journal*, 18: 3–19.

Bernstein, R.S., P.J. Baxter, H. Falk, R. Ing, L. Foster, and F. Frost, 1986. Immediate public health concerns and actions in volcanic eruptions: Lessons from the Mount St. Helens eruptions, May 18–October 16, 1980. *American Journal of Public Health*, 76,(Suppl.): 25–37.

Brawley, O.W., and S.T. Barnes, 2001. Potential agents for prostate cancer chemoprevention. *Epidemiologic Reviews*, 23: 168–172.

Brooks, J.P., 2004. Biological Aerosols Generated from the Land Application of Biosolids: Microbial Risk Assessment. Unpublished Ph.D. dissertation, University of Arizona, Tucson.

Brooks, J.P., B.D. Tanner, C.P. Gerba, C.N. Haas, and I.L. Pepper, 2005a. Estimation of Bioaerosol Risk of Infection to Residents Adjacent to a Land Applied Biosolids Site Using an Empirically Derived Transport Model. *Journal of Applied Microbiology*, 98: 397–405.

Brooks, J.P., B.D. Tanner, K.L. Josephson, C.P. Gerba, and I.L. Pepper, 2005b. A National Study on the Incidence of Biological Aerosols from the Land Application of Biosolids: Microbial Risk Assessment. *Journal of Applied Microbiology*, 99: 310–322.

Brownstein, J.S., T.R. Holford, and D. Fish, 2003. A climate-based model predicts the spatial distribution of the Lyme disease vector *Ixodes scapularis* in the United States. *Environmental Health Perspectives*, 111: 1152–1157.

Brunekreef, B., G. Hoek, P. Fischer, and F.T.M. Spieksma, 2000. Relation between airborne pollen concentrations and daily cardiovascular and respiratory-disease mortality. *Lancet*, 355: 1517–1518.

Brusseau, M.L., and G.R. Tick, 2006. Subsurface Pollution. Pp. 259–278 *in* I.L. Pepper, C.P. Gerba, and M.L. Brusseau (eds.), *Environmental and Pollution Science, 2nd Edition.* San Diego, Calif., Elsevier Science/Academic Press, 532 pp.

Brusseau, M.L., C.M. McColl, G. Famisan, and J.F. Artiola, 2006. Chemical Contaminants. Pp. 132–143 *in* I.L. Pepper, C.P. Gerba, and M.L. Brusseau (eds.), *Environmental and Pollution Science, 2nd Edition.* San Diego, Calif., Elsevier Science/Academic Press, 532 pp.

Bryant, E., 1991. *Natural Hazards.* New York, Cambridge University Press, 294 pp.

Brys, M., A.D. Nawrocka, E. Miekos, C. Zydek, M. Fokinski, A. Barecki, and W.M. Krajewska, 1998. Zinc and cadmium analysis in human prostate neoplasms. *Biological Trace Element Research,* 59: 145–152.

Burd, G.I., D.G. Dixon, and B.R. Glick, 2000. Plant growth-promoting bacteria that decrease heavy metal toxicity in plants. *Canadian Journal of Microbiology,* 46: 237–245.

Burton, I., R.W. Kates, and G.F. White, 1978. *The Environment as Hazard.* New York, Oxford University Press, 240 pp.

Camus, M., J. Siemiatycki, and B. Meek, 1998. Nonoccupational exposures to chrysotile asbestos and the risk of lung cancer. *New England Journal of Medicine,* 338: 1565–1571.

Cannon, H. L., G.G. Connally, J.B. Epstein, J.G. Parker, I. Thornton, and B.G. Wixson, 1978. Rocks: The geologic source of most trace elements. Pp. 17-31 *in* National Research Council, *Distribution of Trace Elements Related to Occurrence of Certain Cancers, Cardiovascular Diseases, and Urolithiasis.* Washington, D.C., National Academy of Sciences, 200 pp.

Cataldo, D.A., and R.E. Wildung, 1978. Soil and plant factors influencing the accumulation of heavy metals by plants. *Environmental Health Perspectives,* 27: 149–159.

CDC (Centers for Disease Control and Prevention), 1999. Achievements in Public Health, 1900–1999: Fluoridation of Drinking Water to Prevent Dental Caries. *Morbidity and Mortality Weekly Report,* 48: 933–940.

CDC (Centers for Disease Control and Prevention), 2002. Surveillance for Asthma—United States, 1980–1999. *Morbidity and Mortality Weekly Report,* 51: 1–13.

CDC (Centers for Disease Control and Prevention), 2004. Lyme Disease—United States, 2001–2002. *Morbidity and Mortality Weekly Report,* 53: 365–369.

CDC (Centers for Disease Control and Prevention), 2005. *Third National Report on Human Exposure to Environmental Chemicals.* Atlanta, Ga., National Center for Environmental Health, 467 pp.

Centeno, J.A., 2000. The diversity of trace elements and toxic metal ions in environmental health and disease. Notes, Metals, Health and Environment, Centre for Continuing Education. New Zealand, Christchurch.

Centeno, J.A., M.A. Gray, F.G. Mullick, P.B. Tchounwou, and C.H. Tseng, 2005. Arsenic in Drinking Water and Health Effects. Pp. 415–439 *in* T.A. Moore, A. Black, J.A. Centeno, J. Harding, and D. Trumm (eds.), *Metal Contaminants in New Zealand—Sources, Treatments, and Effects on Ecology and Human Health.* Christchurch, New Zealand, Resolutionz Press, 490 pp.

CEQ (Council on Environmental Quality), 2006. Environmental Quality Statistics, Section 5: Air Quality Tables 5.5 and 5.6. Available online at *http://ceq.eh.doe.gov/Nepa/reports/statistics/air.html* accessed May 2006.

Chaney, R.L., 1983. Potential Effects of Waste Constituents on the Food Chain. Pp. 152–240 *in* J.F. Parr, P.B. Marsh, and J.M. Kla (eds.), *Land Treatment of Hazardous Waste.* Park Ridge, N.J., Noyes Data Corp.

Chapelle, F.H., 2001. *Ground-Water Microbiology and Geochemistry, 2nd ed.* New York, John Wiley & Sons.

Chesworth, W., 2002. Sustainability and the end of history. *Geotimes,* 47: 5, 52.

Chou, Y-J., N. Huang, C-H. Lee, S-L. Tsai, J-H. Tsay, L-S. Chen, and P. Chou, 2003. Suicides after the 1999 Taiwan earthquake. *International Journal of Epidemiology*, 32: 1007–1014.

Chou, Y-J. N. Huang, C-H. Lee, S-L. Tsai, L-S. Chen, and H-J. Chang, 2004. Who is at risk of death in an earthquake? *American Journal of Epidemiology*, 160: 688–695.

Christian, W.J., C. Hopenhayn, J.A. Centeno, and T.I. Todorov, 2006. Distribution of urinary selenium and arsenic among pregnant women exposed to arsenic in drinking water. *Environmental Research*, 100: 115–122.

Chua, K.B., K.J. Goh, K.T. Wong, A. Kamarulzaman, P.S.K. Tan, T.G. Ksiazek, S.R. Zaki, G. Paul, S.K. Lam, and C.T. Tan, 1999. Fatal encephalitis due to Nipah virus among pig-farmers in Malaysia. *Lancet*, 354: 1257–1259.

Clark, L.C., K. Cantor, and W.H. Allaway, 1991. Selenium in forage crops and cancer mortality in US counties. *Archives of Environmental Health*, 46: 37–42.

Clark, L.C., G.F. Combs, and B.W. Turnbull, 1996. Effects of selenium supplementations for cancer prevention in patients with carcinoma of the skin. *Journal of the American Medical Association*, 276: 1957–1963.

Clarkson, T.W., 2002. The three modern faces of mercury. *Environmental Health Perspectives*, 110 (Suppl. 1): 11–23.

Clinkenbeard, J.P., R.K. Churchill, and K. Lee, 2002. Guidelines for geologic investigations of naturally occurring asbestos in California. *California Geological Survey, Special Publication*, 124, 70 pp.

Cohen, B.J., 2005. *Memmler's The Human Body in Health and Disease*. New York, Lippincott, Williams, and Wilkins, 577 pp.

Combs, G.F., Jr., 2005. Geological Impacts on Nutrition. Pp. 161–177 *in* O. Selinus, B.J. Alloway, J.A. Centeno, R.B. Finkelman, R. Fuge, U. Lindh, and P. Smedley (eds.), *Essentials of Medical Geology*. London, Elsevier Academic Press, 812 pp.

Combs, G.F., Jr., and S.B. Combs, 1984. The nutritional biochemistry of selenium. *Annual Review of Nutrition*, 4: 257-280.

Corona-Cruz, A., G. Gold-Bouchot, M. Gutierrez-Rojas, O. Monroy-Hermosillo, and E. Favela, 1999. Anaerobic-aerobic biodegradation of DDT (dichlorodiphenyl trichloroethane) in soils. *Bulletin of Environmental Contamination and Toxicology*, 63: 219–225.

Cortinas, M.R., M.A. Guerra, C.J. Jones, and U. Kitron, 2002. Detection, characterization, and prediction of tick-borne disease foci. *International Journal of Medical Microbiology*, 291(Suppl. 33): 11–20.

Costello, L.C., and R.B. Franklin, 1998. Novel role of zinc in the regulation of prostate citrate metabolism and its implications in prostate cancer. *Prostate*, 35: 285–296.

Cromley, E.K., 2003. GIS and disease. *Annual Review of Public Health*, 24: 7–24.

Cromley, E.K., and S.L. McLafferty, 2002. *GIS and Public Health*. New York, Guilford Press.

Delange, F., B. de Benoist, E. Pretell, and J.T. Dunn, 2001. Iodine deficiency in the world: Where do we stand at the turn of the century? *Thyroid*, 11: 437–447.

Derbyshire, E., X.M. Meng, and R.A. Kemp, 1998. Provenance, transport and characteristics of modern Aeolian dust in western Gansu Province, China, and interpretation of the Quaternary loess record. *Journal of Arid Environments*, 39: 497–516.

Dingman, L., 2002. *Physical Hydrology, 2nd Edition*. Upper Saddle River, N.J., Prentice Hall, 646 pp.

Doran, J.W., and J.T. Sims, 2002. Sustaining earth and its people. *Geotimes*, 47: 5.

Edmunds, M., and P. Smedley, 2005. Fluoride in Natural Waters. Pp. 301-329 *in* O. Selinus, B.J. Alloway, J.A. Centeno, R.B. Finkelman, R. Fuge, U. Lindh, and P. Smedley (eds.), *Essentials of Medical Geology*. London, Elsevier Academic Press, 812 pp.

Eganhouse, R.P., M.J. Baedecker, I.M. Cozzarelli, G.R. Aiken, K.A. Thorn, and T.F. Dorsey, 1993. Crude oil in a shallow sand and gravel aquifer, II. Organic Geochemistry. *Applied Geochemistry*, 8: 551–567.

Ehrlich, H.L., 1996. *Geomicrobiology, 3rd edition*. New York, Marcel Dekker.

Elliott, P., J.C. Wakefield, N.G. Best, and D.J. Briggs, 2000. Spatial epidemiology: methods and applications. Pp. 3–14 *in* P. Elliott, J.C. Wakefield, N.G. Best, and D.J. Briggs (eds.), *Spatial Epidemiology: Methods and Applications*. Oxford, Oxford University Press.

Ennemoser, O., P. Ambach, P. Brunner, P. Schneider, W. Oberaigner, F. Purtscheller, V. Stingl, and G. Keller., 1994. Unusually high indoor radon concentrations from a giant rock slide. *The Science of the Total Environment*, 151: 235–240.

EPA (Environmental Protection Agency), 1997. *The Benefits and Costs of the Clean Air Act, 1970 to 1990*. Available online at *http://yosemite.epa.gov/ee/epa/eermfile.nsf/vwAN/EE-0295-1.pdf/$File/EE-0295-1.pdf/*, accessed May 2006.

EPA (Environmental Protection Agency), 1998. *National Air Quality and Emissions Trends Report, 1998*. Available online at *http://www.epa.gov/air/airtrends/aqtrnd98/*, accessed May 2006.

EPA (Environmental Protection Agency), 1999. *Regulatory Impact Analysis - Control of Air Pollution from New Motor Vehicles: Tier 2 Motor Vehicle Emissions Standards and Gasoline Sulfur Control Requirements*. Available online at *http://www.epa.gov/tier2/finalrule.htm*, accessed December 2006.

EPA (Environmental Protection Agency), 2003. *List of Drinking Water Contaminants and MCLS*. EPA916-F-03-016. Office of Water, U.S. Environmental Protection Agency, Washington, D.C.

EPA (Environmental Protection Agency), 2006a. *Fact Sheet: Final Revisions to the National Ambient Air Quality Standards for Particle Pollution (Particulate Matter)*. Available online at *http://www.epa.gov/oar/particlepollution/pdfs/20060921_factsheet.pdf*, accessed September 2006.

EPA (Environmental Protection Agency), 2006b. *MINTEQA2*. Available online at *http://www.epa.gov/ceampubl/mmedia/minteq/index.htm* accessed October 2006.

Ernst, W.G., 1990. *The Dynamic Planet*. New York, Columbia University Press, 280 pp.

Evans, W.C., G.W. Kling, M.L. Tuttle, G. Tanyileke, and L.D. White, 1993. Gas buildup in Lake Nyos, Cameroon, the recharge process and its consequences. *Applied Geochemistry*, 8: 207–221.

Falco, R.C., and D. Fish, 1988a. Prevalence of *Ixodes dammini* near the homes of Lyme disease patients in Westchester County, New York. *American Journal of Epidemiology*, 127: 826–830.

Falco, R.C., and D. Fish, 1988b. Ticks parasitizing humans in a Lyme disease endemic area of southern New York State. *American Journal of Epidemiology*, 128: 1146–1152.

Falco, R.C., T.J. Daniels, and D. Fish, 1995. Increase in abundance of immature *Ixodes scapularis* (Acari: Ixodidae) in an emergent Lyme disease endemic area. *Journal of Medical Entomology*, 32: 522–526.

Fankhauser, R.L., J.S. Noel, S.S. Monroe, T. Ando, and R.I. Glass, 1998. Molecular epidemiology of "Norwalk-like viruses" in outbreaks of gastroenteritis in the United States. *Journal of Infectious Diseases*, 178: 1571–1578.

Fields, S., 2004. Global nitrogen: Cycling out of control. *Environmental Health Perspectives*, 112: A557–A563.

Filippelli, G.M., M.A.S. Laidlaw, J.C. Latimer, and R. Raftis, 2005. Urban lead poisoning and medical geology: An unfinished story. *GSA Today*, 15: 4–11.

Finkelman, R.B., H.C.W. Skinner, G.S. Plumlee, and J.E. Bunnell, 2001. Medical Geology. *Geotimes*, 46: 20–23.

Fish, D., 1995. Environmental risk and prevention of Lyme disease. *American Journal of Medicine*, 98: 2–9.

Fish, D., 1996. Remote sensing and landscape epidemiology. *Advances in the Astronautical Sciences*, 91: 1057–1063.

Fish, D., and C. Howard, 1999. Methods used for creating a national Lyme disease risk map. *Morbidity and Mortality Weekly Report*, 48: 21–22.

Fleet, J.C., 1997. Dietary selenium repletion may reduce cancer incidence in people at high risk who live in areas with low soil selenium. *Nutrition Review*, 55: 277–279.

Freeth, F.T., and R.R.F. Kay, 1987. Lake Nyos gas disaster. *Nature*, 325: 104–105.

Friedman, G.M., 1988. *Radon in the Northeast: Perspectives and Geologic Research*. Troy, N.Y., Northeastern Environmental Science.

Galloway, J.N., J.D. Aber, J.W. Erisman, S.P. Seitzinger, R.W. Howarth, E.B. Cowling, and B.J. Cosby, 2003. The nitrogen cascade. *BioScience*, 53: 341–356.

Gates, A.E., and L.C.S. Gundersen (eds.), 1992. *Geologic Controls on Radon*. Special Paper, Geological Society of America, Boulder, Colo.

Gerba, C.P., and I.L. Pepper, 2006. Microbial Contaminants. Pp. 144–168 *in* I.L. Pepper, C.P. Gerba, and M.L. Brusseau (eds.), *Environmental and Pollution Science, 2nd Edition*. San Diego, Calif., Elsevier Science/Academic Press, 532 pp.

Gerba, C.P., K.A. Reynolds, and I.L. Pepper, 2006. Drinking Water Treatment and Water Security. Pp. 468–486 *in* I.L. Pepper, C.P. Gerba, and M.L. Brusseau (eds.), *Environmental and Pollution Science, 2nd Edition*. San Diego, Calif., Elsevier Science/Academic Press, 532 pp.

Gibson, R.S., A.L. Heath, M.L. Limbaga, N. Prosser, and C.M. Skeaff, 2001. Are changes in food consumption patterns associated with lower biochemical zinc status among women from Dunedin, New Zealand? *British Journal of Nutrition*, 86: 71–80.

Gihring, T.M., and J.F. Banfield, 2001. Arsenite oxidation and arsenate respiration by a new *Thermus* isolate. *FEMS Microbiology Letters*, 204: 335–340.

Ginsburg, J.M., and W.D. Lotspeich, 1963. Interrelations of arsenate and phosphate transport in the dog kidney. *American Journal of Physiology*, 205: 707–714.

Glass, G.E., B.S. Schwartz, J.M. Morgan III, D.T. Johnson, P.M. Noy, and E. Israel, 1995. Environmental risk factors for Lyme disease identified with geographic information systems. *American Journal of Public Health*, 85: 944–948.

Gleick, P.H., 1998. *The World's Water 1998–1999*. Washington, D.C., Island Press.

Goldman, L., and D. Ausiello, 2004. *Cecil Textbook of Medicine, 22nd Edition*. London, Elsevier Academic Press, 2,656 pp.

Goodchild, M.F., 1992. Geographical information science. *International Journal of Geographical Information Systems*, 6: 31–45.

Grattan, J., R. Rabartin, S. Self, and T. Thordarson, 2005. Volcanic air pollution and mortality in France, 1783–1784. *Comtes Rendus Geoscience*, 337: 641–651.

Graves, B. (ed.), 1987. *Radon, Radium and Other Radioactivity in Groundwater*. Chelsea, Mich., Lewis Publications.

Gregus, Z., A. Gyurasics, and L. Koszorus, 1998. Interactions between selenium and group Va-metalloids (arsenic, antioimony, and bismuth) in the biliary excretion. *Environmental Toxicology and Pharmacology*, 5: 89–99.

Griffin, S.O., K. Jones, and S.L. Tomar, 2001. An economic evaluation of community water fluoridation. *Journal of Public Health Dentistry*, 61: 78–86.

Griffiths, W.D., and G.A.L. DeCosemo, 1994. The assessment of bioaerosols: A critical review. *Journal of Aerosol Science*, 25: 1425–1458.

Guerra, M., E. Walker, C. Jones, S. Paskewitz, M.R. Cortinas, A. Stancil, L. Beck, M. Bobo, and U. Kitron, 2002. Predicting the risk of Lyme disease: Habitat suitability for *Ixodes scapularis* in the north central United States. *Emerging Infectious Diseases*, 8: 289–297.

Gunatilaka, A.A.L., 2006. Natural products from plant associated microorganisms. Distribution: Structural diversity, bioactivity and implications of their occurrence. *Journal of Natural Products*, 69: 509–526.

Guptill, S.C., and C.G. Moore, 2005. Investigating Vector-borne and Zoonotic Diseases with Remote-sensing and GIS. Pp. 645–665 *in* O. Selinus, B.J. Alloway, J.A. Centeno, R.B. Finkelman, R. Fuge, U. Lindh, and P. Smedley (eds.), *Essentials of Medical Geology*. London, Elsevier Academic Press, 812 pp.

Halloran, M.E., 2001. Concepts of Transmission and Dynamics. Pp. 56-85 *in* J.C. Thomas and D.J. Weber, (eds.), *Epidemiologic Methods for the Study of Infectious Diseases*. Oxford, Oxford University Press, 496 pp.

Harrison, P., and F. Pearce, 2000. *AAAS Atlas of Population & Environment*: American Association for the Advancement of Science. Berkeley, University of California Press, 204 pp.

Hartman, T.J., D. Albanes, P. Pietinen, A.M. Hartman, M. Rautalahti, J.A. Tangrea, and P.R. Taylor, 1998. The association between baseline vitamin E, selenium, and prostate cancer in the alpha-tocopherol, beta-carotene Cancer Prevention Study. *Cancer Epidemiology Biomarkers and Prevention*, 7: 335–340.

Hay, S.I., J. Cox, D.J. Rogers, S.E. Randolph, D.I. Stern, G.D. Shanks, M.F. Myers, and R.W. Snow, 2002. Climate change and the resurgence of malaria in the East African highlands. *Nature*, 415: 905–909.

Heinke, G.W., 1996. Microbiology and epidemiology. Pp 254-302 *in* J.G. Henry and G.W. Heinke. *Environmental Science and Engineering. 2nd edition.* Upper Saddle River, N.J., Prentice Hall, 778 pp.

Hem, J.D., 1985. *Study and Interpretation of the Chemical Characteristics of Natural Water, 3rd edition.* USGS Water Supply Paper, 2254, 151 pp.

Hewitt, K., 1997. *Regions of Risk: A Geographical Introduction to Disasters.* Edinburgh, Addison Wesley Longman Ltd.

HHS (U.S. Department of Health and Human Services), 1991. *Review of Fluoride: Benefits and Risks.* Available online at *http://www.health.gov/environment/ReviewofFluoride/default.htm*, accessed January 2007.

HHS (U.S. Department of Health and Human Services), 2006. *The Health Insurance Portability and Accountability Act of 1996.* Available online at *http://www.cms.hhs.gov/HIPAAGenInfo/02_TheHIPAALawandRelated%20Information.asp#TopOfPage*, accessed November 2006.

Highwood, E.J., and D.S. Stevenson, 2003. Atmospheric impact of the 1783–1784 Laki eruption: Part II. Climatic effect of sulphate aerosol. *Atmospheric Chemistry and Physics*, 3: 1177–1189.

Hobson, D.L., 1997. *Clarification of the Health Information Portability and Accountability Act—Extension of Remarks by the Honorable David L. Hobson in the House of Representatives.* Available online at *http://www.cms.hhs.gov/HIPAAGenInfo/02_TheHIPAALawandRelated%20Information.asp#TopOfPage*, accessed November 2006.

Holmes, C.W., and R. Miller, 2004. Atmospherically transported elements and deposition in the southeastern United States: local or transoceanic? *Applied Geochemistry*, 19: 1189–1200.

Hooda, P.S., C.J.K. Henry, T.A. Seyoum, L.D.M. Armstrong, and M.B. Fowler, 2004. The potential impact of soil ingestion on human mineral nutrition. *Science of the Total Environment*, 333: 75–87.

Horwell, C.J., R.J. Sparks, T.S. Brewer, E.W. Lewellin, and B.J. Williamson, 2003. Characterization of respirable volcanic ash from the Soufriere Hills volcano, Montserrat, with implications for human health standards. *Bulletin of Volcanology*, 65: 346–362.

Hsueh, Y.M., Y.F. Ko, Y.K. Huang, H.W. Chen, H.Y. Chiou, Y.L. Huang, M.S. Yang, and C.J. Chen, 2003. Determinants of inorganic arsenic methylation capability among residents of the Lanyang Basin, Taiwan—Arsenic, selenium exposure and alcohol consumption. *Toxicology Letters*, 137: 49–63.

Huang, Q., W. Chen, and X. Guo, 2004. Immobilization and species of heavy metals in soils in the absence and presence of Rhizobia. *Soil Science and Plant Nutrition*, 50: 935–939.

Innerarity, S., 2000. Hypo-magnesemia in acute and chronically ill patients. *Critical Care Nursing Quarterly*, 23: 1–19.

IOM (Institute of Medicine), 2000. *Clearing the Air: Asthma and Indoor Air Exposures*. Washington, D.C., National Academy Press, 438 pp.

IOM (Institute of Medicine), 2006. *Asbestos: Selected Cancers*. Washington, D.C., The National Academies Press, 328 pp.

IPCC (Intergovernmental Panel on Climate Change), 2001a. *Climate Change 2001: The Scientific Basis. Contribution of Working Group I to the Third Assessment Report of the Intergovernmental Panel on Climate Change*, J.T. Houghton, Y. Ding, D.J. Griggs, M. Noguer, P.J. van der Linden, X. Dai, K. Maskell, and C.A. Johnson (eds.). Cambridge, England, Cambridge University Press, 881 pp.

IPCC (Intergovernmental Panel on Climate Change), 2001b. *Climate Change 2001: Impacts, Adaptation, and Vulnerability: Contribution of Working Group II to the Third Assessment Report of the Intergovernmental Panel on Climate Change*. Cambridge, England, Cambridge University Press, 1032 pp.

Islam, F.S., A.G. Gault, C. Boothman, D.A. Polya, J.M. Charnock, D. Chatterjee, and J.R. Lloyd, 2004. Role of metal-reducing bacteria in arsenic release from Bengal delta sediments. *Nature (London)*, 430: 68–71.

IWMI (International Water Management Institute), 2005. Notes on preliminary tsunami damage rapid assessment for coastal wetlands in Sri Lanka (south coast). Available online at *http://www.recoverlanka.net/data/IWMI_Sri_Lanka_Tsunami_Wetland.doc*, accessed April 2005.

Jackson, C.R., and S.L. Dugas, 2003. Phylogenetic analysis of bacterial and archaeal arsC gene sequences suggests an ancient, common origin for arsenate reductase. *BMC Evolutionary Biology*, 18: 1–10.

Jackson, C.R., E.F. Jackson, S.L. Dugas, K. Gamble, and S.E. Williams, 2003. Microbial transformations of arsenite and arsenate in natural environments. *Recent Research Developments in Microbiology*, 7: 103–118.

Johnson, W., and J. Paone, 1982. *Land Utilization and Reclamation in the Mining Industry, 1930–1980*. U.S. Department of the Interior, Bureau of Mines Information Circular 8862, Washington, D.C.

Jolley, R.L., R.J. Bull, W.P. Davis, S. Katz, M.H. Roberts Jr., and V.A. Jacobs (eds.), 1984. *Water Chlorination: Chemistry, Environmental Impact and Health Effects*, Vol. 5. Chelsea, Mich., Lewis Publishers.

Kabata-Pendias, A., and H.S. Pendias, 2001. *Trace Elements in Soils and Plants, 3rd edition*. Boca Raton, Fla., CRC Press.

Kalipeni, E., and J. Oppong, 1998. The refugee crisis in Africa and implications for health and disease: A political ecology approach. *Social Science and Medicine*, 46: 1637–1653.

Kellogg, C.A., D.W. Griffin, V.H. Garrison, K.K. Peak, N. Royall, R.R. Smith, and E.A. Shinn, 2004. Characterization of aerosolized bacteria and fungi from desert dust events in Mali, West Africa. *Aerobiologia*, 20: 99–110.

Kierstein, S., F.R. Poulain, Y. Cao, M. Grous, R. Mathias, G. Kierstein, M.F. Beers, M. Salmon, R.A. Panettieri, and A. Haczku, 2006. Susceptibility to ozone-induced airway inflammation is associated with decreased levels of surfactant protein D. *Respiratory Research*, 7, 9 pp.

Kilburn, K.H., and R.H. Warshaw, 1995. Hydrogen sulfide and reduced-sulfur gases adversely affect neurophysiological functions. *Toxicology and Industrial Health*, 11: 185–197.

Kodama, H., 1999. Introduction to mineralogy of soil environments. *Kobutsugaku Zasshi* (Journal of the Mineralogical Society of Japan), 28: 13–21.

Koenig, J.Q., 1999. *Health Effects of Ambient Air Pollution*. Boston, Kluwer Academic Publishers.

Kolpin, D.W., E.T. Furlong, M.T. Meyer, E.M. Thurman, S.D. Zaugg, L.B. Barber, and H.T. Buxton, 2002. Pharmaceuticals, hormones, and other organic wastewater contaminants in U.S. streams, 1999–2000: A national reconnaissance. *Environmental Science and Technology*, 36: 1202–1211.

Konikow, L.F., and P.D. Glynn, 2005. Modeling Groundwater Flow and Quality. Pp. 737–768 *in* O. Selinus, B.J. Alloway, J.A. Centeno, R.B. Finkelman, R. Fuge, U. Lindh, and P. Smedley (eds.), *Essentials of Medical Geology*. London, Elsevier Academic Press, 812 pp.

Krieger, N., P. Waterman, J.T. Chen, M.J. Soobader, S.V. Subramanian, and R. Carson, 2002. Zip code caveat: Bias due to spatiotemporal mismatches between zip codes and U.S. census-defined geographic areas—The Public Health Disparities Geocoding Project. *American Journal of Public Health*, 92: 1100–1102.

Kumar, K., S.C. Gupta, S.K. Baidoo, Y. Chandler, and C.J. Rosen, 2005. Antibiotic uptake by plants from soil fertilized with animal manure. *Journal of Environmental Quality*, 34: 2082–2085.

Lam, S.K., and K.B. Chua, 2002. Nipah virus encephalitis outbreak in Malaysia. *Clinical Infectious Diseases*, 34(Suppl. 2): S48–S51.

Lee, S-Z., H.E. Allen, C.P. Huang, D.S. Sparks, P.F. Sanders, and W.J.G.M. Peijnenburg, 1996. Predicting soil-water partition coefficients for cadmium. *Environmental Science and Technology*, 30: 3418–3424.

Legator, M.S., C.R. Singleton, D.L. Morris, and D.L. Philips, 2001. Health effects from chronic low-level exposure to hydrogen sulfide. *Archives of Environmental Health*, 56: 123–131.

Leor, J., W.K. Poole, and R.A. Kloner, 1996. Sudden cardiac death triggered by an earthquake. *New England Journal of Medicine*, 334: 413–419.

Lerman, B.B., N. Ali, and D. Green, 1980. Megaloblastic dyserythropoietic anemia following arsenic ingestion. *Annals of Clinical and Laboratory Science*, 10: 515–517.

Liang, J.Y., Y.Y. Liu, J. Zou, R.B. Franklin, L.C. Costello, and P. Feng, 1999. Inhibitory effect of zinc on human prostatic carcinoma cell growth. *Prostate*, 40: 200–207.

Liddell, F.D., 1997. Magic, menace, myth and malice. *Annals of Occupational Hygiene*, 41: 3–12.

Lighthart, B., and L.D. Stetzenbach, 1994. Distribution of Microbial Bioaerosols. Pp. 68–98 *in* B. Lighthart and A.J. Mohr (eds.), *Atmospheric Microbial Aerosols, Theory and Applications*. New York, Chapman & Hall.

Lofts, S., and E. Tipping, 1998. An assemblage model for cationic binding by natural particulate matter. *Geochimica et Cosmochimica Acta*, 62: 2609–2625.

Lovley, D.R., 1987. Organic matter mineralization with reduction of ferric iron: A review. *Geomicrobiology Journal*, 5: 375–399.

Mace, R.E., R.S. Fisher, D.M. Welch, and S.P. Parra, 1997. *Extent, Mass, and Duration of Hydrocarbon Plumes from Leaking Petroleum Storage Tank Sites in Texas*. University of Texas, Bureau of Economic Geology, Circular GC97-1.

Magnaval, J.-F., L.T. Glickman, P. Dorchies, and B. Morassin, 2001. Highlights of human toxocariasis. *Korean Journal of Parasitology*, 39: 1–11.

Maier, R.M., I.L. Pepper, and C.P. Gerba, 2000. *Environmental Microbiology*. San Diego, Calif., Academic Press.

Mannino, D., D. Homa, C. Pertowski, A. Ashizawa, L. Nixon, C. Johnson, L. Ball, E. Jack, and D. Kang, 1998. Surveillance for Asthma Prevalence—United States, 1960–1995. *MMWR Morbidity and Mortality Weekly Report*, 47(SS-1): 1–28.

Matthias, A.D., 2005. Monitoring Near-Surface Air Quality. Pp. 164–181 *in* J.F. Artiola, I.L. Pepper, and M. Brusseau (eds.), *Environmental Monitoring and Characterization*. San Diego, Calif., Elsevier Academic Press.

Matthias, A.D., S.A. Musil, and H.L. Bohn, 2006. Physical-Chemical Characteristics of the Atmosphere. Pp. 46–57 *in* I.L. Pepper, C.P. Gerba, and M.L. Brusseau (eds.), *Environmental and Pollution Science, 2nd Edition*. San Diego, Calif., Elsevier Science/Academic Press, 532 pp.

McElroy, M.B., 2002. *Atmospheric Environment: Effect of Human Activity*. Princeton, N.J., Princeton University Press.

McGee, K.A., and T.M. Gerlach, 1998. Annual cycle of magmatic CO2 in a tree-kill soil at Mammoth Mountain, California: Implications for soil acidification. Geology, 26: 463–466.

McIlroy, A., 2001. *SOS for Canada's H2O*. The Guardian Unlimited. Available online at *http://www.guardian.co.uk/elsewhere/journalist/story/0,7792,487205,00.html*, accessed September 2006.

McMichael, A.J., 2002. Population, environment, disease, and survival: Past patterns, uncertain futures. *Lancet*, 359: 1145–1148.

Mead, P.S., L. Slutsker, V. Dietz, L.F. McCaig, J.S. Bresee, C. Shapiro, P.M. Griffin, and R.V. Tauxe, 1999. Food-related illness and death in the United States. *Emerging Infectious Diseases*, 5: 607–625.

Melnick, A.L., 2001. *Introduction to Geographic Information Systems for Public Health*. New York, Aspen.

Methel, O., 2003. Role of lipopolysaccharide (LPS) in asthma and other pulmonary conditions. *Journal of Endotoxin Research*, 9: 293–300.

Mielke, H.W., C. Gonzales, E. Powell, S. Coty, and A. Shah, 2003. Anthropogenic Distribution of Lead. Pp. 119–124 *in* H.C.W. Skinner and A.R. Berger (eds.), *Geology and Health: Closing the Gap*. New York, Oxford University Press, 179 pp.

Mileti, D.S., 1999. *Disasters by Design: A Reassessment of Natural Hazards in the United States*. Washington, D.C., Joseph Henry Press, 351 pp.

Morens, D.M., G.K. Folkers, and A.S. Fauci, 2004. The challenge of emerging and re-emerging infectious diseases. *Nature*, 430: 242–249.

Moynahan, E.J., 1979. Trace elements in man. *Philosophical Transactions of the Royal Society of London. Series B, Biological Sciences*, B288: 65–79.

Munich Re Group, 2000. *Topics 2000: Natural Catastrophes—The Current Position*. Available online at *http://www.munichre.com/publications/302-02354_en.pdf*, accessed March 2004.

Nadelman, R.B., and G.P. Wormser, 2005. Poly-ticks: Blue state versus red state for Lyme disease. *Lancet*, 365: 280.

NASA (National Aeronautics and Space Administration), 2001. *Microbes and the Dust They Ride in Pose Potential Health Risks*. Earth Observatory Release No. 01-120, NASA, Washington, D.C.

Nelson, M.A., B.W. Porterfield, E.T. Jacobs, and L.C. Clark, 1999. Selenium and prostate cancer prevention. *Seminars in Urology and Oncology*, 17: 91–96.

NETL (National Energy Technology Laboratory), 2006. *Osage-Skiatook Petroleum Environmental Research Project*. Available online at *http://www.netl.doe.gov/technologies/oil-gas/Petroleum/projects/Environmental/Produced_Water/15238.htm*, accessed May 2006.

NIEHS (National Institute of Environmental Health Sciences), 2006. *Particles: Size Makes All the Difference*. Environmental Health Perspectives, NIEHS. Available online at *http://www.ehponline.org/science-ed/2006/particle.pdf*, accessed December 2006.

NIH–NSF (National Institutes of Health and National Science Foundation), 2005. Review of the Joint National Institutes of Health–National Science Foundation Ecology of Infectious Diseases Program. Available online at *http://www.fic.nih.gov/about/eid_review2005.pdf*, accessed May 2006.

Njemanze, P.C., J. Anozie, J.O. Ihenacho, M.J. Russell, and A.B. Uwaeziozi, 1999. Application of risk analysis and geographic information system technologies to the prevention of diarrheal diseases in Nigeria. *American Journal of Tropical Medicine and Hygiene*, 61: 356–360.

NOHSC (National Occupational Health and Safety Commission), 1999. Australian Mesothelioma Register Report 1999.

Noji, E.K. (ed.), 1997. *The Public Health Consequences of Disasters.* New York, Oxford University Press, 468 pp.

Noji, E.K., 2005. Disasters: Introduction and state of the art. *Epidemiologic Reviews*, 27: 3–8.

Norris, F., M. Friedman, P. Watson, C. Byrne, E. Diaz, and K. Kaniasty, 2002. 60,000 disaster victims speak: Part I: An empirical review of the empirical literature, 1981–2001. *Psychiatry*, 65: 206–239.

Norris, F.H., A.D. Murphy, C.K. Baker, and J.L. Perilla, 2004. Postdisaster PTSD over four waves of a panel study of Mexico's 1999 flood. *Journal of Traumatic Stress*, 17: 283–292.

NRC (National Research Council), 1974. *Geochemistry of the Environment: Volume I, The Relation of Selected Trace Elements to Health and Disease.* Washington D.C., National Academy of Sciences, 533 pp.

NRC (National Research Council), 1977. *Geochemistry of the Environment: Volume II, The Relation of Other Selected Trace Elements to Health and Disease.* Washington D.C., National Academy of Sciences, 163 pp.

NRC (National Research Council), 1978. *Geochemistry of the Environment: Volume III, Distribution of Trace Elements Related to the Occurrence of Certain Cancers, Cardiovascular Dieases, and Urolithiasis.* Washington D.C., National Academy of Sciences, 200 pp.

NRC (National Research Council), 1979. *Geochemistry of Water in Relation to Cardiovascular Disease.* Washington D.C., National Academy of Sciences, 98 pp.

NRC (National Research Council), 1981. *Aging and the Geochemical Environment.* Washington D.C., National Academy of Sciences, 141 pp.

NRC (National Research Council), 1987. *Drinking Water and Health, Volume 7, Disinfectants and Disinfectant By-Products.* Washington D.C., National Academy of Sciences, 212 pp.

NRC (National Research Council), 1993. *In Situ Bioremediation: When Does it Work?* Washington D.C., National Academy Press, 224 pp.

NRC (National Research Council), 1995. *Nitrate and Nitrite in Drinking Water.* Washington D.C., National Academy Press, 64 pp.

NRC (National Research Council), 1996. *Mineral Resources and Sustainability: Challenges for Earth Scientists.* Washington, D.C., National Academy Press, 26 pp.

NRC (National Research Council), 1997. *Dietary Reference Intakes for Calcium, Phosphorus, Magnesium, Vitamin D, and Fluoride.* Washington, D.C., National Academy Press, 432 pp.

NRC (National Research Council), 1999a. Health Effects of Exposure to Radon: BEIR VI. Washington, D.C., National Academy Press, 500 pp.

NRC (National Research Council), 1999b. *Hormonally Active Agents in the Environment.* Washington, D.C., National Academy Press, 452 pp.

NRC (National Research Council), 1999c. *Risk Assessment of Radon in Drinking Water.* Washington, D.C., National Academy Press, 432 pp.

NRC (National Research Council), 1999d. *From Monsoons to Microbes: Understanding the Ocean's Role in Human Health.* Washington, D.C., National Academy Press, 144 pp.

NRC (National Research Council), 1999e. *Arsenic in Drinking Water.* Washington, D.C., National Academy Press, 330 pp.

NRC (National Research Council), 2000a. *Natural Attenuation for Groundwater Remediation.* Washington D.C., National Academy Press, 274 pp.

NRC (National Research Council), 2000b. *Clean Coastal Waters: Understanding and Reducing the Effects of Nutrient Pollution.* Washington D.C., National Academy Press, 428 pp.

NRC (National Research Council), 2000c. *Toxicological Effects of Methylmercury.* Washington D.C., National Academy Press, 368 pp.

NRC (National Research Council), 2001a. *Under the Weather: Climate, Ecosystems, and Infectious Disease.* Washington D.C., The National Academies Press, 160 pp.

NRC (National Research Council), 2001b. *Arsenic in Drinking Water: 2001 Update.* Washington, D.C., National Academy Press, 244 pp.

NRC (National Research Council), 2002a. *Biosolids Applied to Land: Advancing Standards and Practices.* Washington, D.C., The National Academies Press, 346 pp.

NRC (National Research Council), 2002b. *Estimating the Public Health Benefits of Proposed Air Pollution Regulations.* Washington, D.C., The National Academies Press, 170 pp.

NRC (National Research Council), 2002c. *Malaria Control During Mass Population Movements and Natural Disasters.* Washington, D.C., The National Academies Press, 164 pp.

NRC (National Research Council), 2004a. *Groundwater Fluxes Across Interfaces.* Washington D.C., The National Academies Press, 100 pp.

NRC (National Research Council), 2004b. *Confronting the Nation's Water Problems: The Role of Research.* Washington D.C., The National Academies Press, 324 pp.

NRC (National Research Council), 2004c. *From Source Water to Drinking Water: Workshop Summary.* Washington D.C., The National Academies Press, 126 pp.

NRC (National Research Council), 2004d. *Facilitating Interdisciplinary Research.* Washington D.C., The National Academies Press, 306 pp.

NRC (National Research Council), 2005. *The Geological Record of Ecological Dynamics: Understanding the Biotic Effects of Future Environmental Change.* Washington D.C., The National Academies Press, 200 pp.

NRC (National Research Council), 2006a. *Fluoride in Drinking Water: A Scientific Review of EPA's Standards.* Washington, D.C., The National Academies Press, 530 pp.

NRC (National Research Council), 2006b. *Managing Coal Combustion Residues in Mines.* Washington D.C., The National Academies Press, 228 pp.

NRC (National Research Council), 2006c. *Beyond Mapping: Meeting National Needs Through Enhanced Geographic Information Science.* Washington D.C., The National Academies Press, 100 pp.

NTP (National Toxicology Program), 2005. Report on Carcinogens, Eleventh Edition. U.S. Department of Health and Human Services, Public Health Service, National Toxicology Program. Available online at *http://ntp-server.niehs.nih.gov/index.cfm?objectid=32BA9724-F1F6-975E-7FCE50709CB4C932*, accessed December 2006.

Nuckols, J.R., M.H. Ward, and L. Jarup, 2004. Using Geographical Information Systems for Exposure Assessment in Environmental Epidemiology Studies. *Environmental Health Perspectives,* 112: 1007–1015.

Ogawa, K., I. Tsuji, K. Shiono, and S. Hisamichi, 2000. Increased acute myocardial infarction mortality following the 1995 Great Hanshin-Awaji earthquake in Japan. *International Journal of Epidemiology,* 29: 449–455.

Ogunlewe, J.O., and D.N. Osegbe, 1989. Zinc and cadmium concentrations in indigenous blacks with normal, hypertrophic, and malignant prostates. *Cancer,* 63: 1388–1392.

Ohlendorf, H.M., R.L. Hothem, C.M. Bunck, and K.C. Marois, 1990. Bioaccumulation of selenium in birds at Kesterson Reservoir, California. *Archives of Environmental Contamination and Toxicology,* 19: 495–507.

Ong, C.G., M.J. Herbel, R.A. Dahlgren, and K.K. Tanji, 1997. Trace elements (Se As Mo B) contamination of evaporites in hypersaline agricultural evaporation ponds. *Environmental Science and Technology*, 31: 831–836.

Oremland, R.S., and J.F. Stolz, 2003. The ecology of arsenic. *Science*, 300: 939–944.

Oremland, R.S., S.E. Hoeft, J.M. Santini, N. Bano, R.A. Hollibaugh, and J.T. Hollibaugh, 2002. Anaerobic oxidation of arsenite in Mono Lake water and by a facultative, arsenite-oxidizing chemoautotroph, strain MLHE-1. *Applied and Environmental Microbiology*, 68: 4795–4802.

Oster, J.D., and S.R. Grattan, 2002. Drainage water reuse. *Irrigation and Drainage Systems*, 16: 297–302.

Ostfeld, S.R., and F. Keesing, 2000. Biodiversity and disease risk: The case of Lyme disease. *Conservation Biology*, 14: 722–728.

Ozen, S., and A. Sir, 2004. Frequency of PTSD in a group of search and rescue workers two months after 2003 Bingol (Turkey) earthquake. *The Journal of Nervous and Mental Disease*, 192: 573–575.

Parkhurst, D.L., and C.A.J. Appelo, 1999. User's guide to PHREEQC (version 2)—a computer program for speciation, batch-reaction, one-dimensional transport, and inverse geochemical calculations. *U.S. Geological Survey Water-Resources Investigations Report*, 312 pp.

Pascual, M., J.A. Ahumada, L.F. Chaves, X. Rodó, and M. Bouma, 2006. Malaria resurgence in the East African highlands: Temperature trends revisited. *Proceedings of the National Academy of Sciences USA*, 103: 5829–5834.

Patz, J.A., T.K. Graczyk, N. Geller, and A.Y. Vittor, 2000. Effects of environmental change on emerging parasitic diseases. *International Journal on Parasitology*, 30: 1395–1405.

Patz, J.A., M. Hulme, C. Rosenzweig, T.D. Mitchell, R.A. Goldberg, A.K. Githeko, S. Lele, A.J. McMichael, and D. Le Sueur, 2002. Climate change: Regional warming and malaria resurgence. *Nature*, 420: 627–628.

Pepper, I.L., and M.L. Brusseau, 2006. Physical-Chemical Characteristics of Soils and the Subsurface. Pp. 13–23 *in* I.L. Pepper, C.P. Gerba, and M.L. Brusseau (eds.), *Environmental and Pollution Science, 2nd Edition*. San Diego, Calif., Elsevier Science/Academic Press, 532 pp.

Pepper, I.L., C.P. Gerba, and M.L. Brusseau (eds.), 2006. *Environmental and Pollution Science, 2nd Edition*. San Diego, Calif., Elsevier Science/Academic Press, 532 pp.

Peters, S.C., and J.D. Blum, 2003. The source and transport of arsenic in a bedrock aquifer, New Hampshire, USA. *Applied Geochemistry*, 18: 1773–1787.

Pew (The Pew Environmental Health Commission), 2000. *American's Environmental Health Gap: Why the Country Needs a Nationwide Health Tracking Network*. Available online at *http://www.earthscape.org/pmain/sites/pehc.html*, accessed October 2006.

Plant, J.A., and D.L. Davis, 2003. Breast and Prostate Cancer: Sources and Pathways of Endocrine-Disrupting Chemicals (EDCs). Pp. 95–98 *in* H.C.W. Skinner and A.R. Berger (eds.), *Geology and Health: Closing the Gap*. New York, Oxford University Press, 179 pp.

Platt, R.H., 1999. *Disasters and Democracy: The Politics of Extreme Natural Events*. Washington, D.C., Island Press.

Platz, E.A., and K.J. Helzlsouer, 2001. Selenium, zinc, and prostate cancer. *Epidemiologic Reviews*, 23: 93–101.

Plumlee, G.S., R.A. Morton, T.P. Boyle, J.H. Medlin, and J.A. Centeno, 2000. *An Overview of Mining-Related Environmental and Human Health Issues, Marinduque Island, Philippines: Observations from a Joint U.S. Geological Survey–Armed Forces Institute of Pathology Reconnaissance Field Evaluation, May 12–19, 2000*. U. S. Geological Survey Open-File Report 00-397. Available online at *http://pubs.usgs.gov/of/2000/ofr-00-0397/ofr-00-0397.pdf*, accessed April 2006.

Poulstrup, A., and H.L. Hansen, 2004. Use of GIS and exposure modeling as tools in a study of cancer incidence in a population exposed to airborne dioxin. *Environmental Health Perspectives,* 112: 1032–1036.

Press, C. M., J.E. Loper, and J.W. Kloepper, 2001. Role of iron in rhizobacteria-mediated induced systemic resistance of cucumber. *Phytopathology,* 91: 593–598.

Prospero, J.M., 2001. African dust in America. *Geotimes,* 46: 24–27.

Queirolo, F., S. Stegan, M. Restovic, M. Paz, P. Ostapczuk, M. J. Schwuger, and L. Munoz, 2000. Total arsenic, lead, and cadmium levels in vegetables cultivated at the Andean villages of northern Chile. *Science of the Total Environment,* 255: 75–84.

Radovanovic, H., and A.A. Koelmans, 1998. Prediction of in situ trace metal distribution coefficients for suspended solids in natural waters. *Environmental Science and Technology,* 32: 753–759.

Ragnarsdottir, K.V., 2000. Environmental fate and toxicology of organophosphate pesticides. *Journal of the Geological Society of London,* 157(Part 4): 859–876.

Rahman, M.H., M.M. Rahman, C. Watanabe, and K. Yamamoto, 2003. Arsenic Contamination of Groundwater in Bangladesh and Its Remedial Measures. Pp. 9–22 *in Arsenic Contamination in Groundwater—Technical and Policy Dimensions.* Tokyo, United Nations University, 44 pp.

Rajaretnam, G., and H.B. Spitz, 2000. Effect of leachability on environmental risk assessment for naturally occurring radioactive materials in petroleum oil fields. *Health Physics,* 78: 191–198.

Rasmussen, P.E., 1996. *Trace Metals in the Environment: A Geological Perspective.* Geological Survey of Canada, Bulletin 429, 26 pp.

Rayman, M.P., 2005. Selenium in cancer prevention: A review of the evidence and mechanism of action. *Proceedings of the Nutrition Society,* 64: 527–542.

Reeves, P.G., and R.L. Chaney, 2002. Nutritional status affects the absorption and whole-body and organ retention of cadmium in rats fed rice-based diets. *Environmental Science and Technology,* 36: 2684–2692.

Ricketts, T.C., 2003. GIS and public health. *Annual Review of Public Health,* 24: 1–6.

Robbins, E.I., and M. Harthill, 2003. Life in a Copper Province. Pp. 105–112 *in* H.C.W. Skinner and A.R. Berger, (eds.), *Geology and Health: Closing the Gap.* New York, Oxford University Press, 179 pp.

Roehrborn, C.G., 2000. Acute relief or future prevention: Is urology ready for preventive health care? *Urology,* 56: 12–19.

Rogers, J.R., 2000. Nutrient-Driven Colonization and Weathering of Silicates. Unpublished Ph.D. dissertation, University of Texas at Austin.

Rogers, J.R., and P.C. Bennett, 2004. Mineral stimulation of subsurface microorganisms: Release of limiting nutrients from silicates. *Chemical Geology,* 203: 91–108.

Rogers, J.R., P.C. Bennett, and W.J. Choi, 1998a. Feldspars as a source of nutrients for microorganisms. *American Mineralogist,* 83: 1532–1540.

Rogers, J.R., P.C. Bennett, and W.J. Ullman, 1998b. Biochemical release of a limiting nutrient from feldspars. Proceedings of the V.M. Goldschmidt Conference 1998 (Toulouse, France), *Mineralogical Magazine,* 62A: 1283–1284.

Root, T.L., J.M. Bahr, and M.B. Gotkowitz, 2005. Geochemical and environmental controls on arsenic in groundwater near Lake Geneva, Wisconsin. Pp. 161–174 *in* P.A. O'Day, D. Vlassopoulos, X. Meng, and L.G. Benning (eds.), *Advances in Arsenic Research: Integration of Experimental and Observational Studies and Implications for Mitigation.* ACS Symposium Series, vol. 915, American Chemical Society.

Rose, J.B., P.R. Epstein, E.K. Lipp, B.H. Sherman, S.M. Bernard, and J.A. Patz, 2001. Climate variability and change in the United States: Potential impacts on water- and foodborne diseases caused by microbiologic agents. *Environmental Health Perspectives*, 109(Suppl. 2): 211–221.

Rosenlund, M., N. Berglind, J. Hallqvist, T. Bellander, and G. Bluhm, 2005. Daily intake of magnesium and calcium from drinking water in relation to myocardial infarction. *Epidemiology*, 16: 570–576.

Rosner, D., and G. Markowitz, 1991. *Deadly Dust Silicosis and the Politics of Occupational Disease in Twentieth-Century America.* Princeton, N.J., Princeton University Press.

Ross, M., and R.P. Nolan, 2003. History of asbestos discovery and use and asbestos related disease in context with the occurrence of asbestos within ophiolites complexes. Pp. 447–470 *in* Special Paper 373. Geological Society of America, Boulder, Colo.

Rude, R.K., 1998. Clinical review. Magnesium deficiency: A cause of heterogeneous disease in humans. *Journal of Bone Mineral Research*, 13: 749–758.

Rushton, G., 2003. Public health, GIS, and spatial analytic tools. *Annual Review of Public Health*, 24: 43–56.

Rutta, E., R. Kipingili, H. Lukonge, S. Assefa, E. Mitsilale, and S. Rwechungura, 2001. Treatment outcome among Rwandan and Burundian refugees with sputum smear-positive tuberculosis in Ngara, Tanzania. *International Journal of Tuberculosis and Lung Diseases*, 5: 628–632.

Ryker, S.J., 2001. Mapping arsenic in groundwater. *Geotimes*, 46: 34–36.

Salama, P., and T.J. Dondero, 2001. HIV surveillance in complex emergencies. *AIDS*, S4–S12.

Santamaria, J., and G.A. Toranzos, 2003. Enteric pathogens and soil: A short review. *International Microbiology*, 6: 5–9.

Saris, N.E., E. Mervaala, H. Karppanen, J.A. Khawaja, and A. Lewenstam, 2000. Magnesium. An update on physiological, clinical and analytical aspects. *Clinica Chimica Acta* 294: 1–26.

Saxe, J.K., C.A. Impellitteri, W.J.G.M. Peijnenburg, and H.E. Allen, 2001. A novel model describing heavy metal concentrations in the earthworm, *Eisenia andrei*. *Environmental Science and Technology*, 35: 4522–4529.

SCF (Scientific Committee for Food), 1993. *Nutrient and Energy Intakes for the European Community.* Reports of the Scientific Committee for Food, 31st Series. European Commission, Luxembourg.

Schecher, W.D., and D.L. McAvoy, 1998. *MINEQL+: A Chemical Equilibrium Modeling System, User's Manual, Version 4.0.* Hallowell, Maine, Environmental Research Software.

Schmidt, K.A., and R.S. Ostfeld, 2001. Biodiversity and the dilution effect in disease ecology. *Ecology*, 82: 609–619

Searl, A., A. Nicholl, and P.J. Baxter, 2002. Assessment of the exposure of islanders to ash from the Soufriere Hills volcano, Montserrat, Brutusgh West Indies. *Occupational and Environmental Medicine*, 39: 523–531.

Seiler, R.L., S.D. Zaugg, J.M. Thomas, and D.L. Howcroft, 1999. Caffeine and pharmaceuticals as indicators of wastewater contamination in wells. *Ground Water*, 37: 405–410.

Selinus, O., B.J. Alloway, J.A. Centeno, R.B. Finkelman, R. Fuge, U. Lindh, and P. Smedley (eds.), 2005. *Essentials of Medical Geology.* London, Elsevier Academic Press, 812 pp.

Shamberger, R.J., and C.E. Willis, 1969. Selenium distribution and human cancer mortality. *CRC Critical Reviews in Clinical Laboratory Science*, 2: 211–221.

Sharif, E.N., J. Douwes, P.H.M. Hoet, G. Doekes, and B. Nemery, 2004. Concentrations of domestic mite and ret allergens and endotoxin I palestine. *Allergy*, 59: 623–631.

Sherer, Y., R. Bitzur, H. Cohen, A. Shaish, D.Varon, Y. Shoenfeld, and D. Harats, 2001. Mechanisms of action of the antiatherogenic effect of magnesium: Lessons from a mouse model. *Magnesium Research*, 14: 173–179.

Silver, S., and D. Keach, 1982. Energy-dependent arsenate flux: The mechanism of plasmid-mediated resistance. *Proceedings of the National Academy of Sciences USA*, 79: 6114–6118.
Skinner, H.C.W., and A.R. Berger (eds.), 2003. *Geology and Health: Closing the Gap*. New York, Oxford University Press, 179 pp.
Skinner, H.C.W., M. Ross, and C. Frondel, 1988. *Asbestos and Other Fibrous Materials: Mineralogy, Crystal Chemistry and Health Effects*. New York, Oxford University Press, 204 pp.
Smedley, P., and D.G. Kinniburgh, 2005. Arsenic in Groundwater and the Environment. Pp. 263–299 *in* O. Selinus, B.J. Alloway, J.A. Centeno, R.B. Finkelman, R. Fuge, U. Lindh, and P. Smedley (eds.), *Essentials of Medical Geology*. London, Elsevier Academic Press, 812 pp.
Smith, K., 2001. *Environmental Hazards: Assessing Risk and Reducing Disaster*. London, Routledge.
Sobsey, M.D., 1989. Inactivation of health-related microorganisms in water by disinfection processes. *Water Science and Technology*, 21: 179–195.
Steere, A.C., S.E. Malawista, D.R. Snydman, R.E. Shope, W.A. Andiman, M.R. Ross, and F.M. Steele, 1977. Lyme arthritis: An epidemic of oligoarticular arthritis in children and adults in three Connecticut communities. *Arthritis & Rheumatism*, 20: 7–17.
Stetzenbach, L.D., 2001. Introduction to Aerobiology. Pp 801–813 *in* C.J. Hurst, R.L. Crawford, G.R. Knudsen, M.J. McInerny, and L.D. Stetzenbach (eds.), *Manual of Environmental Microbiology, 2nd Ed.* Washington, D.C., ASM Press.
Stirzaker, R.J., J.B. Passioura, and Y. Wilms, 1996. Soil structure and plant growth: Impact of bulk density and biopores. *Plant and Soil*, 185: 151–162.
Stone, R., 2004. Iceland's doomsday scenario. *Science*, 306: 1278–1281.
Straub, T.M., I.L. Pepper, and C.P. Gerba, 1992. Persistence of viruses in desert soils amended with anaerobically digested sewage sludge. *Applied Environmental Microbiology*, 58: 636–641.
Strobel, G., and B. Daisy, 2003. Bioprospecting for microbial endophytes and their natural products. *Microbiology and Molecular Biology Reviews*, 67: 491–502.
Stumm, W., 1990. *Chemistry of the Solid-Water Interface—Processes at the Mineral-Water and Particle-Water Interface in Natural Systems*. New York, John Wiley & Sons.
Stumm, W., and J. Morgan, 1996. *Aquatic Chemistry: Chemical Equilibria and Rates in Natural Waters*. New York, John Wiley & Sons.
Swan, S.H., K.M. Main, F. Liu, S.L. Stewart, R.L. Kruse, A.M. Calafat, C.S. Mao, J.B. Redmon, C.L. Ternand, S. Sullivan, J.L. Teague, and the Study for Future Families Research Team, 2005. Decrease in anogenital distance among male infants with prenatal phthalate exposure. *Environmental Health Perspectives*, 113: 1056–1061.
Tanser, F.C., B. Sharp, and D. le Sueur, 2003. Potential effect of climate change on malaria transmission in Africa. *Lancet*, 362: 972–978.
Tate, R.L., 2000. *Soil Microbiology*. New York, John Wiley, 508 pp.
Taylor, L.H., S.M. Latham, and M.E. Woolhouse, 2001. Risk factors for human disease emergence. *Philosophical Transactions of the Royal Society of London. Series B, Biological Sciences*, 356: 983–989.
Thrush, J., 2000. Cadmium in the New Zealand ecosystem. Unpublished Ph.D. dissertation. Otago University, Dunedin, New Zealand.
Tipping, E., 1994. WHAM—A chemical equilibrium model and computer code for waters, sediments, and soils incorporating a discrete site/electrostatic model of ion-binding by humic substances. *Computers and Geosciences*, 21: 973–1023.
Tipping, E., 1998. Humic Ion-Binding Model VI: An improved description of the interactions of protons and metal ions with humic substances. *Aquatic Geochemistry*, 4: 3–48.
Tipping, E., 2002. *Cation Binding by Humic Substances*. Cambridge, Cambridge University Press.

Toole, M.J., 1997. Communicable Diseases and Disease Control. Pp. 79–100 *in* E.K. Noji (ed.), *The Public Health Consequences of Disasters*. New York, Oxford University Press, 468 pp.

Torres, O., P.K. Bhartia, J.R. Herman, and Z. Ahmad, 1998. Derivation of aerosol properties from satellite meaurements of backscattered ultraviolet radiation, theoretical basis. *Journal of Geophysical Research*, 103: 17,099–17,110.

Torres, O., P.K. Bhartia, A. Sinyuk, and P. Ginoux, 2002. A long-term record of aerosol optical depth from TOMS observations and comparison to Aeronet measurements. *Journal of Atmospheric Science*, 59: 398–413.

Torsvik, V., J. Goksøyr, and F.L. Daae, 1990. High diversity of DNA of soil bacteria. *Applied and Environmental Microbiology*, 56: 782–787.

Townsend, A.R., R.W. Howarth, F.A. Bazzaz, M.S. Booth, C.C. Cleveland, S.K. Collinge, A.P. Dobson, P.R. Epstein, E.A. Holland, D.R. Keeney, M.A. Mallin, C.A. Rogers, P. Wayne, and A.H. Wolfe, 2003. Human health effects of a changing global nitrogen cycle. *Frontiers in Ecology and the Environment*, 1: 240–246.

Tsai, C.H., F.W. Lung, and S.Y. Wang, 2004. The 1999 Ji-Ji (Taiwan) earthquake as a trigger for acute myocardial infarction. *Psychosomatics*, 45: 477–482.

UNAIDS (Joint United Nations Programme on HIV/AIDS), 2005. *Strategies to Support the HIV-Related Needs of Refugees and Host Populations*. Available online at *http://data.unaids.org/publications/irc-pub06/JC1157-Refugees_en.pdf*, accessed October 2006.

UNDP (United Nations Development Programme), 2003. *Disease, HIV/AIDS, and Capacity Limitations: A Case of the Public Agriculture Sector in Zambia*. New York, United Nations Development Programme.

UNDP (United Nations Development Programme), 2004. *Reducing Disaster Risk: A Challenge for Development*. New York, United Nations Development Programme.

USGS (United States Geological Survey), 2000. *Invisible CO2 Gas Killing Trees at Mammoth Mountain, California*. USGS Fact Sheet 172-96, version 2. Available online at *http://pubs.usgs.gov/fs/fs172-96/*, accessed April 2006.

Vedel, S., 1995. *Health Effects of Inhalable Particles: Implications for British Columbia*. Air Resources Branch, Ministry of Environment, Lands and Parks, Victoria.

Ver Ploeg, M., and E. Perrin (eds.), 2004. *Eliminating Health Disparities: Measurement and Data Needs*. Washington, D.C., The National Academies Press, 294 pp.

Vineis, P., 2004. A self-fulfilling prophecy: Are we underestimating the role of the environment in gene-environment interaction research? *International Journal of Epidemiology*, 33: 945–946.

Vogt, T.M., R.G. Ziegler, B.I. Graubard, C.A. Swanson, R.S. Greenberg, J.B. Schoenberg, M.G. Swanson, R.B. Hayes, and S.T. Mayne, 2003. Serum selenium and risk of prostate cancer in US blacks and whites. *International Journal of Cancer*, 103: 664–670.

Walworth, J.L., 2005. Physical Contaminants. Pp. 281–296 *in* J.F. Artiola, I.L. Pepper, and M.L. Brusseau (eds.), *Environmental Monitoring and Characterization*. San Diego, Calif., Elsevier Academic Press.

Warwick, R., and P.L. Williams (eds.), 1973. *Gray's Anatomy*. Philadelphia, W.B. Saunders.

Wasley, A., 1995. Epidemiology in the disaster setting. *Current Issues in Public Health*, 1: 131–135.

WCED (World Commission on Environment and Development), 1987. *Our Common Future*. World Commission on Environment and Development (Brundtland Commission Report). New York, Oxford University Press.

Webber, M.M., 1985. Selenium prevents the growth stimulatory effects of cadmium on human prostatic epithelium. *Biochemical and Biophysical Research Communications*, 127: 871–877.

Weigle, K.A., C. Santrich, F. Martinez, L. Valderrama, and N.G. Saravia, 1993. Epidemiology of cutaneous leishmaniasis in Colombia: Environmental and behavioral risk factors for infection, clinical manifestations, and pathogenicity. *Journal of Infectious Diseases*, 168: 709–714.

WHO (World Health Organization), 2001. *Arsenic in Drinking Water. Fact Sheet No. 210*, Revised May 2001. Available online at *http://www.who.int/mediacentre/factsheets/fs210/en/index.html*, accessed February 2006.

WHO (World Health Organization), 2004. *Some Drinking-Water Disinfectants and Contaminants, Including Arsenic.* International Agency for Research on Cancer. IARC Monographs on the Evaluation of Carcinogenic Risks to Humans, 84. Lyon, France.

WHO (World Health Organization) and UNICEF (United Nations Children's Fund), 2000. *Global Water Supply and Sanitation Assessment Report.* Available online at *http://www.who.int/docstore/water_sanitation_health/Globassessment/GlobalTOC.htm*, accessed October 2006.

Wilkins, C.H., and S.J. Birge, 2005. Prevention of osteoporotic fractures in the elderly. *American Journal of Medicine*, 118: 1190–1195.

Williams, J.H., T.D. Phillips, P.E. Jolly, J.K. Stiles, C.M. Jolly, and D. Aggarwal, 2004. Human Aflatoxicosis in Developing Countries: A Review of Toxicology, Exposure, Potential Health Consequences, and Interventions. *American Journal of Clinical Nutrition*, 80: 1106–1122.

Wilson, R., and J.D. Spengler (eds.), 1996. *Particles in Our Air: Concentrations and Health Effects*. Cambridge, Mass., Harvard University Press, 265 pp.

Wolfe, N.D., M.N. Eitel, J. Gockowski, P.K. Muchaal, C. Nolte, A.T. Prosser, J.N. Torimiro, S.F. Weise, and D.S. Burke, 2000. Deforestation, hunting and the ecology of microbial emergence. *Global Change and Human Health*, 1: 10–25.

Wolman, M.G., 2002. The human impact: Some observations. *Proceedings of the American Philosophical Society*, 146: 81–98.

Wylie, A.G., K.F. Bailey, J.W. Kelse, and R.J. Lee, 1993. The importance of width in asbestos fibers carcinogenicity and its implications for public policy. *American Industrial Hygiene Association Journal*, 54: 239–252.

Young, P., 1997. Major microbial diversity initiative recommended. *ASM News*, 63: 417–421.

Zaleski, K.J., K.L. Josephson, C.P. Gerba, and I.L. Pepper, 2005. Survival, growth, and regrowth of enteric indicator and pathogenic bacteria in biosolids, compost, soil, and land applied biosolids. *Journal of Residuals Science and Technology*, 2: 49–63.

Zehnder, G.W., J.F. Murphy, E.J. Sikora, and J.W. Kloepper, 2001. Application of rhizobacteria for induced resistance. *European Journal of Plant Pathology*, 107: 39–50.

Zheng, B.S., X. Yu, J. Zhand, and D. Zhou, 1996. Environmental geochemistry of coal and endemic arsenism in southwest Guizhou Province, China. *30th International Geologic Congress Abstracts*, 3: 410.

Zhou, L., R. Sriram, G.S. Visvesvara, and L. Xiao, 2003. Genetic variations in the internal transcribed spacer and mitochondrial small subunit rRNA gene of *Naegleria* spp. *Journal of Eukaryotic Microbiology*, 50(Suppl. 1): 522–526.

Appendixes

Appendix A

Committee and Staff Biographies

H. Catherine W. Skinner (*Chair*) holds teaching and research positions in the Departments of Geology and Geophysics, Yale University, and Orthopedics and Rehabilitation, Yale Medical School. She previously held positions at the National Institute of Arthritis and Metabolic Diseases, the National Institute of Dental Research, and the Department of Biology, Yale University. Dr. Skinner is a fellow of the Geological Society of America, the American Association for the Advancement of Science, and the Mineralogical Society of America and has served as president of the Connecticut Academy of Arts and Sciences. She is a trustee of the Geological Society of America Foundation. Her research areas include minerals, particularly minerals found in life forms, microbes, invertebrates and vertebrates, and the processes of biomineralization. Dr. Skinner is an author of the major reference text *Dana's New Mineralogy*, three other books, and over 70 research papers. She received her B.A. from Mount Holyoke College, her M.A. from Radcliffe, and her Ph.D. from the University of Adelaide, South Australia.

Herbert E. Allen is a professor of environmental engineering at the University of Delaware. Before joining the faculty of the University of Delaware in 1989, he was director of the Environmental Studies Institute and professor of chemistry at Drexel University; previously he was on the faculty of the Department of Environmental Engineering at the Illinois Institute of Technology. Dr. Allen received his Ph.D. and B.S. from the University of Michigan and his M.S. from Wayne State University. Dr. Allen's research has primarily been concerned with the fate and effects of trace metals in aquatic and soil environments. He has authored more than 170 journal papers, book chapters, and major reports and has edited seven books and prepared numerous reports and proceedings papers. He has been the principal or co-principal investigator for over 70 research projects funded by government and industry. His principal areas of research are the fate of metals in aquatic and terrestrial environments and the development of site-specific criteria. He headed a multiuniversity consortium of universities, supported by the Environmental Protection Agency from 1994 until 2000 that conducted research on the fate and effects of metals and organics in natural water systems, and now heads the newly formed multiuniversity EPA Center for the Study of Metals in the Environment. Dr. Allen was a member of the organizing committee for the 1993 EPA Annapolis workshop and a member of the 1996 SETAC Pellston conference to review water quality criteria for metals. He served as chairman of the Organizing Committee for the Workshop on Metal Speciation that was held in Jekyll Island, Georgia, every two years from 1987 through 1995. He has also served as a consultant to a number of industrial companies, government agencies, and the World Health Organization.

Jean M. Bahr is a professor of hydrogeology in the Department of Geology and Geophysics, University of Wisconsin-Madison, where she has been a faculty member since 1987. She is a former chair of the Water Resources Management Program at UW and a member of the Geological Engineering Program faculty. Her research interests include both naturally occurring and anthropogenic sources of groundwater contamination and the coupled physical and biogeochemical processes responsible for subsurface contaminant transport. She is currently serving a three-year term on the Council of the Geological Society of America and in 2003 was the Geological Society of America's Hydrogeology Division Birdsall-Dreiss Distinguished Lecturer. She earned a B.A. in geology from Yale University and M.S. and Ph.D. degrees in applied earth sciences (hydrogeology) from Stanford University.

Philip C. Bennett is a professor in the Department of Geological Sciences at the University of Texas, Austin. He teaches undergraduate and graduate courses in aqueous geochemistry and hydrology and graduate courses in geomicrobiology, organic geochemistry, and geochemical kinetics. His research is primarily in the area of microbial and environmental geochemistry, mineral weathering kinetics, and geomicrobiology, with a focus on the influence of geochemistry and geology on subsurface microbial ecology. Dr. Bennett has a B.S. from Evergreen State College (Olympia, WA), a M.S. from SUNY Syracuse, and a Ph.D. from Syracuse University.

Kenneth P. Cantor is a senior investigator in the Division of Cancer Epidemiology and Genetics at the National Cancer Institute, where he has directed studies of cancer and environmental factors since 1977. His research interests focus on the epidemiological investigation of cancer risks associated with occupational and environmental factors and their interaction with other exposures and host factors. His particular area of investigation deals with water contaminants (nitrate, arsenic, disinfection byproducts), pesticides, electromagnetic radiation, and the role of genetic factors in susceptibility to these and other factors. Dr. Cantor received his B.A. from Oberlin College, a Ph.D. from the University of California at Berkeley, and an M.P.H. from the Harvard School of Public Health.

José A. Centeno is a senior research scientist and chief of the Division of Biophysical Toxicology, Department of Environmental and Infectious Disease Sciences, U.S. Armed Forces Institute of Pathology in Washington, D.C. He is also director of the International Tissue and Tumor Repository on Chronic Arseniasis, the Registry on Uranium and Depleted Uranium, and the International Registry on Medical Geology. He holds adjunct faculty professorships at major universities and medical institutions, including the George Washington University School of Public Health, Turabo University(Puerto Rico), Jackson State University, and Hope University Medical School. Since 2005 he has served as an officer for the International Union of Geological Sciences and its Commission on Geoscience for Environmental Management. He serves as a founding member and cochairman of the International Medical Geology Association. His research focuses on environmental toxicology, environmental pathology, medical geology, and health effects of trace elements, toxic trace metals, and metalloids, and he has conducted research and taught training activities on medical geology in over 30 countries. Dr. Centeno received his B.S. and M.S. degrees from the University of Puerto Rico and a Ph.D. from Michigan State University.

Lois K. Cohen recently retired as the associate director for international health at the National Institute of Dental and Craniofacial Research, where she directed both the Office of International Health and the World Health Organization Collaborating Center for Dental and Craniofacial Research. She is now a consultant to the NIDCR. Dr. Cohen is a sociologist, specializing in oral health and international collaborative research. She has published numerous research papers and edited four books in the socio-dental sciences, health systems research, and health promotion, and serves on the advisory boards for the PAHO/WHO programs for oral health, the Middle East Center for Dental Education, and the Canadian Institute for Health Research's Institute for Musculoskeletal Health and Arthritis. She also serves on the editorial boards of the *Journal of the American Dental Association* and the *African Journal of Oral Health* and is a peer reviewer for other journals in the social science, medicine, and public health fields. Awarded an honorary doctorate from Purdue University, she has also been named a senior distinguished scientist by the International Association for Dental Research and the D.C. Sociological Society and has been awarded an honorary fellowship in the American Association for the Advancement of Science, the American and International Colleges of Dentistry, the American Dental Association, and the Academy of Dentistry Internationale. The founding president of Behavioral Scientists in Dental Research, she has served as an officer for several public health, dental research, and social science associations. She has held academic appointments in the Department of Sociology and Anthropology at Howard University and in the Department of Social Medicine and Health Policy at Harvard Medical School. Dr. Cohen has a B.A. from the University of Pennsylvania and M.S. and Ph.D. degrees in sociology from Purdue University.

Paul R. Epstein is associate director of the Center for Health and the Global Environment at Harvard Medical School and is a medical doctor trained in tropical public health. Dr. Epstein has worked in medical, teaching, and research capacities in Africa, Asia and Latin America and in 1993 coordinated an eight-part series on health and climate change for the British medical journal, *Lancet*. Dr. Epstein coedited the report *Climate Change Futures: Health, Ecological and Economic Dimensions*, with support from Swiss Re and the United Nations Development Programme. He has worked with the Intergovernmental Panel on Climate Change, the National Oceanic and Atmospheric Administration, and the National Aeronautics and Space Administration to assess the health impacts of climate change and develop health applications of climate forecasting and remote sensing.

W. Gary Ernst (NAS) is Benjamin M. Page Professor Emeritus in the Department of Geological and Environmental Sciences at Stanford University. He was previously professor of geology and geophysics at University of California, Los Angeles, chairman of the Department of Geology (1970–1974), chairman of the Department of Earth and Space Sciences (1978–1982), and UCLA director of the Institute of Geophysics and Planetary Physics (1987–1989). In 1989 he moved to Stanford for a five-year term as dean of the School of Earth Sciences. Dr. Ernst's research encompasses the physical chemistry of rocks and minerals; Phanerozoic interactions of lithospheric plates and mobile mountain belts, especially in central Asia, the Circumpacific, and the western Alps; early Precambrian petrotectonic evolution; ultrahigh-pressure subduction-zone metamorphism and tectonics; geobotanical studies of arid mountains; and earth system science/remote sensing. Dr. Ernst is a trustee of the Carnegie Institution of Washington, D.C. (1990–present) and a fellow of the American Academy of Arts and Sciences, the American Philosophical Society, the American Geophysical Union, the American Association for the Advancement of Science (chair section on geology and geography, 2001), the Geological Society of America (president, 1985–1986; Penrose Medal, 2004), and the Mineralogical Society of America (president, 1980–1981; Roebling Medal, 2006).

Shelley A. Hearne is executive director of Trust for America's Health and a visiting scholar at the Johns Hopkins University Bloomberg School of Public Health, where she teaches public health infrastructure, policy, and advocacy. She was formerly executive director of the Pew Environmental Health Commission at Johns Hopkins. Dr. Hearne is the immediate past-chair of the American Public Health Association's Executive Board, and she has served on many national organizations, including as vice president of the Council on Education for Public Health. Dr. Hearne has previously worked as a program officer at the Pew Charitable Trusts, as acting director of the New Jersey Department of Environmental Protection's Office of Pollution Prevention, and as a research scientist with the Natural Resources Defense Council. She holds a B.A. degree in chemistry and environmental studies with honors from Bowdoin College and a doctorate in environmental health sciences from Columbia University's School of Public Health.

Jonathan D. Mayer is professor of epidemiology, international health, and geography at the University of Washington, with major interests in infectious disease epidemiology and environmental epidemiology. He has served on several National Institutes of Health committees, as well as the Institute of Medicine–National Research Council Committee on Climate, Ecosystems, Infectious Diseases, and Human Health. He is also an affiliate of the Center for Studies in Demography and Ecology, an affiliate of the International Health Program, and a member of the clinical staff of the Tropical Medicine and Infectious Disease Service of the University of Washington Medical Center. He is an elected member of the American College of Epidemiology. Dr. Mayer's specialties are infectious diseases, society, and environmental change; disease ecology; and HIV in sub-Saharan Africa. He is also director of UW's Undergraduate Program in Public Health. His research focuses on the ecology of infectious diseases, global change and climate change and their impact on infectious diseases and population health, and the use of spatial analysis and Geospatial Information Systems in understanding infectious disease patterns. His planned future projects include work in clinical and genetic epidemiology. Dr. Mayer has a B.A. from the University of Rochester and M.A. and Ph.D. degrees from the University of Michigan.

Jonathan Patz is associate professor of environmental studies and population health sciences at the University of Wisconsin-Madison, with a joint appointment with the Nelson Institute for Environmental Studies and the Department of Population Health Sciences. He was formerly director of the Program on Health Effects of Global Environmental Change and assistant professor in the Department of Environmental Health Sciences at the Johns Hopkins University Bloomberg School of Public Health. His research activities are focused on the effects of climate change on head waves, air pollution and water- and vectorborne diseases, and the link between deforestation and resurgent diseases in the Amazon. He was co-chair for the U.S. National Assessment on Climate Variability and Change health sector expert panel and convening lead author for the United Nations/World Bank Millennium Ecosystem Assessment. Dr. Patz holds joint faculty appointments with the Departments of Epidemiology, International Health, Microbiology, Medicine, and Earth and Planetary Sciences, and he is also an affiliate scientist at National Center for Atmospheric Research. He has medical board certification in both occupational/environmental medicine and family medicine and received his medical degree from Case Western Reserve University and his M.P.H. from Johns Hopkins University.

Ian L. Pepper is director of both the Environmental Research Laboratory and the National Science Foundation Water Quality Center, both at the University of Arizona. He is also a professor and research scientist with the Department of Soil, Water, and Environmental Science, and the Department of Microbiology and Immunology at the University of Arizona. Dr. Pepper is an environmental microbiologist specializing in the molecular ecology of the environment. He is a fellow of the Soil Science Society of America, the American Society of Agronomy, and the American Academy of Microbiology. Dr. Pepper has a B.Sc. from the University of Birmingham, Great Britain, and M.S and Ph.D. degrees from Ohio State University.

Liaison from the Board on Health Sciences Policy:

Bernard D. Goldstein (IOM) is dean of the University of Pittsburgh Graduate School of Public Health. Previously he served as director of the Environmental and Occupational Health Sciences Institute, a joint program of Rutgers, the State University of New Jersey and the University of Medicine and Dentistry of New Jersey Robert Wood Johnson Medical School. He was also principal investigator of the Consortium of Risk Evaluation with Stakeholder Participation. Dr. Goldstein was assistant administrator for research and development, Environmental Protection Agency, 1983–1985. His past activities include member and chairman of the NIH Toxicology Study Section and EPA's Clean Air Scientific Advisory Committee and chair of the Institute of Medicine Committee on the Role of the Physician in Occupational and Environmental Medicine, the National Research Council Committees on Biomarkers in Environmental Health Research and Risk Assessment Methodology and the Industry Panel of the World Health Organization Commission on Health and Environment. He is a member of the Institute of Medicine, where he has chaired the Section on Public, Biostatistics, and Epidemiology.

National Research Council Staff

David A. Feary is a senior program officer with the Board on Earth Sciences and Resources, National Research Council. He earned his Ph.D. at the Australian National University before spending 15 years as a research scientist working on continental margin evolution in the marine program at the Australian Geological Survey Organisation. During this time he participated in numerous research cruises—many as chief or co-chief scientist—and most recently was co-chief scientist for the Ocean Drilling Program Leg 182. His research activities have focused on the role of climate as a primary control on carbonate reef formation and developing an improved understanding of cool-water carbonate depositional processes.

Christine M. Coussens is a program officer with the Board on Health Sciences Policy, Institute of Medicine. Currently, she is study director of the Roundtable on Environmental Health Sciences, Research, and Medicine—a neutral environment for key stakeholders in environmental health to gather and discuss areas of mutual concern. She received her Ph.D. in biomedical sciences with a concentration in neuroscience from Northeastern Ohio College of Medicine/Kent State University. As a research fellow at the University of Otago, Dunedin, New Zealand, she coauthored numerous papers on synaptic plasticity and learning and memory. Since joining the IOM, she has worked on reports analyzing national formulary system of the Department of Veterans Affairs and nervous systems disorders in developing countries.

Caetlin M. Ofiesh is a research associate with the Board on Earth Sciences and Resources, National Research Council. Since graduating from

Amherst College with a B.A. in geology, she has worked in the Boulder, Colorado, Office of Environmental Defense and spent a semester teaching geology to high school sophomores in Zermatt, Switzerland. Prior to that, she interned at the American Geological Institute in Alexandria, Virginia, and worked as a research assistant at the University of Virginia's School of Engineering, where she published a paper on global food and water supply issues.

Amanda M. Roberts was a senior program assistant with the Board on Earth Sciences and Resources until August 2006. Before working at the National Academies she interned at the Fund for Peace in Washington, D.C., working on the Human Rights and Business Roundtable. She also worked in Equatorial Guinea, Africa, with the Bioko Biodiversity Protection Program. She is a master's degree student at The Johns Hopkins University in the Environment and Policy program and holds an M.A. in international peace and conflict resolution from Arcadia University, specializing in environmental conflict in sub-Sahara Africa.

Nicholas D. Rogers is a senior program assistant with the Board on Earth Sciences and Resources, National Research Council. He received a B.A. in history, with a focus on the history of science and early American history, from Western Connecticut State University in 2004. He began working for the National Academies in 2006 and has primarily supported the Board on Earth Sciences and Resources on earth resource and geographical science issues.

Appendix B

Acronyms and Abbreviations

ABLES	Adult Blood-Level Epidemiology and Surveillance program
AFIP	Armed Forces Institute of Pathology
ATSDR	Agency for Toxic Substances and Disease Registry
BPH	benign prostatic hyperplasia
BTEX	benzene, toluene, ethylbenzene, and xzylene
CDC	Centers for Disease Control and Prevention
CEC	cation-exchange capacity
CHONSP	carbon, hydrogen, oxygen, nitrogen, sulfur, and phosphorous
COPD	Chronic Obstructive Pulmonary Disease
DDT	dichloro-diphenyl-trichloroethane
DOC	Dissolved Organic Carbon
DoD	Department of Defense
EDCs	endocrine disrupting compounds
EID	Ecology of Infectious Diseases
EPA	Environmental Protection Agency
FIC	Fogarty International Center
GIS	Geographic Information System
GIScience	Geographic Information Science
GUI	Graphical user interface

HIPAA	Health Insurance Portability and Accountability Act
IGERT	Integrative Graduate Education and Research Traineeship
IRB	Institutional Review Board
LOAEC	Lowest Observable Adverse Effect Concentration
LOAEL	Lowest Observable Adverse Effect Level
LPS	lipopolysaccharide
NASA	National Aeronautics and Space Administration
NCEH	National Center for Environmental Health
NCHS	National Center for Health Statistics
NCI	National Cancer Institute
NHANES	National Health and Nutrition Examination Surveys
NHIS	National Health Interview Survey
NIEHS	National Institute of Environmental Health Sciences
NIH	National Institutes of Health
NOAA	National Oceanic and Atmospheric Administration
NOAEC	No Observable Adverse Effect Concentration
NOAEL	No Observable Adverse Effect Level
NRC	National Research Council
NSF	National Science Foundation
PCBs	Polychlorinated biphenyls
PM	particulate matter
PTSD	post-traumatic stress disorder
RFP	Request for Proposal
SBRP	Superfund Basic Research Program
SCAMP	Surface Chemistry Assemblage Model for Particles
SWAMP	Sediment Water Algorithm for Metal Partitioning
TCE	Trichloroethylene
TOMS	Total Ozone Mapping Spectrometer
TRI	Toxics Release Inventory
UNESCO	United Nations Educational, Scientific and Cultural Organization
USGS	United States Geological Survey
UV	ultraviolet light
VOCs	volatile organic compounds